Fluorinated Ionomers

Fluorinated Ionomers

Walther Grot
C. G. Processing, Inc.
Chadd's Ford, Pennsylvania

Plastics Design Library

ISBN: 978-0-8155-1541-8

Library of Congress Cataloging-in-Publication Data
Grot, Walther.
    Fluorinated ionomers / Walther Grot.
        p. cm.
    Includes bibliographical references and index.
    ISBN 978-0-8155-1541-8 (978-0-8155)
1. Ionomers. 2. Organofluorine compounds. 3. Electrolytic cells. I. Title.
    QD382.I45G76   2008
    668.9–dc22   2007035442

Printed in the United States of America
This book is printed on acid-free paper.

10 9 8 7 6 5 4 3 2 1

Published by:
William Andrew Publishing
13 Eaton Avenue
Norwich, NY 13815
1-800-932-7045
www.williamandrew.com

NOTICE

To the best of our knowledge the information in this publication is accurate; however the Publisher does not assume any responsibility or liability for the accuracy or completeness of, or consequences arising from, such information. This book is intended for informational purposes only. Mention of trade names or commercial products does not constitute endorsement or recommendation for their use by the Publisher. Final determination of the suitability of any information or product for any use, and the manner of that use, is the sole responsibility of the user. Anyone intending to rely upon any recommendation of materials or procedures mentioned in this publication should be independently satisfied as to such suitability, and must meet all applicable safety and health standards.

Contents

PDL Fluorocarbon Series Editor's Preface

The original idea for the Fluorocarbon Series was conceived in the mid-1990s. Two important rationales required the development of the collection. First, there were no definitive sources for the study of fluorinated polymers, particularly the commercial products. A researcher seeking the properties and characteristics of fluorinated plastics did not have a single source to use as a reference. Information available from commercial manufacturers of polymers had long been the source of choice. Second, waves of the post-war generation (a.k.a. Baby Boomers) were beginning to retire, thus eroding the available knowledge base in the industry and academia.

The scope of the series has been expanded over time to incorporate other important fluorinated materials. Selection of the topics of the books has been based on the importance of practical applications. Inevitably, a number of fluorinated compounds, important in their own right, have been left out of the series. In each case, the size of its audience has been found simply too small to meet the economic hurdles of publishing.

The first two books of the series cover commercial fluoropolymers (ethylenic); the third book is focused on their applications in the chemical processing industries. The fourth book covers fluoroelastomers, the fifth fluorinated coatings and finishes, and the sixth book is about fluorinated ionomers, such as Nafion®. The seventh handbook represents an extension of the scope of the series to non-polymeric materials. It addresses the preparation, properties, and uses of fluorinated chemicals as refrigerants, fire extinguishers, blowing agents, and cleaning gases. A full list of the books in the PDL Fluorocarbon Series appears at the back of this book.

The authors of the handbooks are leaders in their fields who have devoted their professional careers to acquiring expertise. Each book is a product of decades of each author's experience and research into the available body of knowledge. Our hope is that these efforts will meet the needs of the people who work with fluorinated polymers and chemicals. Future revisions are planned to keep this series abreast of progress in the field.

Sina Ebnesajjad
September 2007

Preface

Fluorinated ionomer polymers are the lesser-known materials, even though they have important industrial applications, particularly in the field of industrial electrochemistry. One of the most important contributions of this polymer to mankind is that it allows elimination of mercury in the production of chlorine and sodium hydroxide, together with a substantial reduction of the electric energy used in this process. The resulting reduction of the overall worldwide electric energy consumption, both industrial and domestic, is estimated to be almost 1%.

This book covers partially fluorinated and perfluorinated polymers containing sufficient ionic groups to dominate the transport properties of the polymer. The emphasis of this book is on the practical aspects of working with fluorinated ionomers. It is intended to help the scientists and engineers in the use of these products, and in the development of new applications and compositions.

The extensive coverage given to perfluorinated ionomers is intentional, because of the practical importance of this group of polymers. Within this group, the emphasis on Nafion® (a trademark of DuPont Company) is not intentional, but due to the extensive coverage that this polymer has received in the literature.

Chapters 1 through 4 present the history, manufacturing and properties of perfluorinated ionomers. Even though Chapters 5 and 6 focus on the applications, the latter is devoted to fuel cell and battery applications of these polymers. Chapters 7 and 8 describe the economics and available commercial membranes of fluorinated ionomers. Chapter 9 provides a list of experimental methods used to characterize perfluorinated ionomers. Chapters 10 through 13 discuss heat sealing and repair, handling and storage, toxicology and safety and suppliers and resources topics. Finally, Chapter 14 offers a glossary of terms and some useful web sites for additional information.

None of the views or information presented in this book reflects the opinions of any of the companies or individuals that have contributed to the book. If there are errors, they are oversights on the part of the authors. A note to the publisher indicating the specific error, for the purpose of correcting future editions, would be much appreciated.

Acknowledgements

I would like to acknowledge the contributions made by many individuals in industry and academia, in particular my coworkers at DuPont, foremost Dr. Frank Gresham, who first conceived the idea of introducing sulfonic acid groups into polytetrafluoro ethylene. The discovery of Nafion® was made possible by a corporate culture of innovation for its own sake, not driven by the needs of the market. When the potential for the two major applications (membranes for chlor-alkali cells and fuel cells) were recognized in the mid-1960s, the development was pushed ahead even though the economics of these applications did not look attractive at that time.

Special thanks are due to many companies that have contributed data and information to this book. Citation has been made where the material appears in the text.

My special thanks also to the Krupp-Uhde Company for the many excellent pictures and diagrams they made available for this book.

This is the first book I wrote, and it would not have been possible without the loving care of my wife Carla during my recovery from a serious operation and the encouragement and help of my editor and friend Dr. Sina Ebnesajjad.

Walther Grot
Chadd's Ford, Pennsylvania
September, 2007

1 Introduction

Fluorinated ionomers, particularly the perfluorinated ionomers developed in the 1960s, have revolutionized the chlor-alkali industry. In this process, the need for the use of hazardous materials such as mercury and asbestos has not only been eliminated, but the economics, particularly with respect to reduced energy consumption, has also been substantially improved. This application has now matured to such an extent that the complete replacement of the two older technologies is only a question of time.

More recently, a new application has emerged in the field of fuel cells. This development is still in flux and is the subject of considerable research in both industry and government institutions. It appears that the full potential of this application is yet to be realized.

The combination of hydrophilic and hydrophobic groups in the same polymer molecule of polymeric fluorinated ionomers results in unique properties and morphologies, which has attracted the attention of industry, researchers, and theoreticians. However, many questions regarding the inner workings of this material remain still unanswered.

1.1 Polymers

Both partially fluorinated and perfluorinated polymers, containing sufficient ionic groups to dominate the transport properties of the polymer, have been described in this book. Ionic groups may include sulfonic and carboxylic groups as well as sulfonamides and sulfonimides. Because of their importance in the synthesis and fabrication of these ionomers, their precursor polymers, containing sulfonyl fluoride or carboxylic ester groups, are also discussed. However, it should be emphasized that these precursor polymers are not ionomers and that they have properties which are quite different from the corresponding ionomers.

The synthesis of a perfluorinated ionomer containing phosphonic acid groups has been reported in the literature [1, 2]. Perfluorinated ionomers containing sulfonyl imide functional groups have also received some attention [3].

Within this broad scope, perfluorinated ionomers containing sulfonic or carboxylic functional groups have been covered most extensively because of their large commercial uses. Within this narrower group, the emphasis has been placed on Nafion®, which has been available for about ten years longer than any of the other competitive materials in its class. DuPont has made both information and samples of Nafion® and its precursor polymer readily available to research groups and commercial users, which has resulted in extensive coverage of Nafion® in the literature.

1.2 Physical Shapes

Most fluorinated ionomers are sold as flat sheets and films, such as extruded or solution cast films, or as composite membranes containing fabric reinforcement added to one or more layers of the ionomer. Extruded capillary tubing is also available. Smaller quantities are sold in the form of pellets for applications such as catalysts or for conversion to liquid compositions. The end-use properties as well as the morphology and structure of these products are discussed in Chapter 4.

References

1. Kato, M., Akiyama, K., Yamabe, M., Reps. Res. Lab., Asahi Glass Co. Ltd., **33**(2), 135, 1983.
2. Kotov, S., Pedersen, S., Qiu, Z., Burton, D., J. Fluorine Chem., **82**, 13–19, 1997.
3. Thomas, B., Shafer, G., Ma, J., Tu, M., DesMarteau, D., J. Fluorine Chem., **125**(8), 1231–1240, 2004.

2 History

The first fluorinated ionomer was discovered in the early 1960s at the DuPont Experimental Station near Wilmington, Delaware [1]. It was a perfluorinated ionomer that later became known as Nafion®. At that time, an exploratory chemistry group within DuPont's Plastics Department headed by Frank Gresham was pursuing a newly discovered synthetic route to prepare perfluorinated vinyl ethers. This route allowed the conversion of almost any perfluorinated acyl fluoride to the corresponding vinyl ether through the addition of hexafluoropropylene epoxide followed by dehalocarbonylation. These perfluoro vinyl ethers were promising monomers for the production of melt-fabricable copolymers of tetrafluoroethylene [2]. Several important monomers, including perfluoro methyl-, ethyl-, and propyl-vinyl ether, became commercially available as result of this work.

One of the vinyl ethers synthesized by this method was based on the reaction product of sulfur trioxide and tetrafluoroethylene. It offered the opportunity to introduce ionic groups into a perfluorinated polymer. Initially, the motivation for this work was simply a curiosity to study polymers with a broad range of compositions, but the presence of ionic groups made this polymer different from anything known before. It was shown that the reactive groups of the precursor polymer allowed vulcanization using curing agents such as magnesium oxide [1]. When combined with perfluoro methyl-vinyl ether as a termonomer, this should yield perfluorinated elastomers with improved properties. Another early expectation was that ionic cross-linking would result in improved mechanical properties, particularly resistance to creep, of fluoropolymers in general. However, none of these early approaches led to any useful products. Instead, the presence of ionic groups had an adverse effect on most of the useful properties of perfluorinated polymers such as unsurpassed dielectric properties, exceptionally low coefficient of friction, and non-stick and hydrophobic behavior. At this point, the identification of commercial uses for this polymer required thinking "outside the box."

The use of Nafion® as a separator membrane in a chlor-alkali cell was demonstrated by Grot in 1964. In 1966, Grot and Selman approached General Electric (GE) regarding the use of this polymer in fuel cells. These two applications weave like supporting threads throughout the entire development of perfluorinated ionomers: While one application was on top and in the limelight, research on the other was taking place in the background in preparation for its move to the top.

While initial experiments indicated that a Nafion® membrane could be used in a chlor-alkali cell, improvements in strength and in hydroxide ion rejection were necessary to meet the needs of the industrial chlor-alkali market. The electrolysis cells developed for the two incumbent technologies (asbestos and mercury) were also not suitable for membrane operation. As a result, the chlor-alkali industry saw little incentive in abandoning the existing technologies, which had been optimized during many decades of development, in favor of a new one, which needed significant improvement in terms of membrane performance and cell design.

Fortunately, at that time there was a critical need for high-performance fuel cell membranes in connection with the space program. The use of Nafion® in this application proved an immediate success. The sales from this program supported a small production project and allowed improvements in monomer synthesis, polymerization processes, and fabrication techniques. In addition, the price of several thousands of dollars per square meter attracted the attention of the management of a department involved in the sale of low-cost products such as polyethylene and Mylar® films.

In the meantime, work on the chlor-alkali application continued, facilitated by the ready availability of starting materials and fabricated shapes. The problem of poor mechanical strength, particularly poor resistance to tear propagation, was solved by the introduction of a reinforcing fabric made of polytetrafluoroethylene [3]. The use of a thin barrier layer, made of a highly selective version of a perfluorinated ionomer, proved to be a powerful and very versatile approach to improved hydroxide ion rejection [4–7]. Two other developments overseas had a major effect on the eventual success of Nafion® in this application:

- In Europe, the introduction of titanium-based anodes (dimensional-stable anodes or DSAs) by Beer [8] and de Nora [9] provided long-term stability of the anode combined with lower cell voltage. The use of these DSAs was synergistic with the use of membrane technology.
- In Japan, the government concluded in 1968 that a mysterious disease that had plagued a local population around the Minamata Bay for more than a decade was caused by the ingestion of fish and shellfish contaminated with methyl mercury. This disease—"Minamata disease"—had caused 46 fatalities in 1956 and was responsible for several thousand cases of serious illness. The source of the mercury was traced back to the discharge of waste streams from the Chisso acetaldehyde plant, which used mercuric sulfate as a catalyst. While it appears that the chlor-alkali industry was not

involved in this environmental disaster, the Japanese government ordered the phase out of the use of mercury in the production of chlorine and caustic soda. As a result, the Japanese industry, with substantial government support, launched a crash program to adopt the newly emerging membrane technology for the manufacture of chlorine and caustic soda.

Asahi Glass's discovery of a barrier layer containing carboxylic acid groups, which provides particularly effective rejection of hydroxyl ions, was an important aspect of this work [10]. The improvement in performance was so large that DuPont developed its own version of a carboxylic barrier layer [11]. Asahi Chemicals was the third company to introduce a similar membrane. Since 1980, these three companies have offered similar membranes for chlor-alkali applications: a main layer consisting of a sulfonic polymer with an imbedded fabric reinforcement that was coated on one surface with a thin barrier layer of carboxylic polymer. The sulfonic polymers used contain fairly long (five or six carbons in addition to two ether groups) linkages between the sulfonic acid group and the polymer backbone.

In 1982, Dow Chemical introduced a short-branch (two carbons plus one ether group) sulfonic polymer (the "Dow polymer") [12, 13]. Although this structure appeared to be superior from a theoretical standpoint, it was never fully commercialized because of the difficulty in synthesizing the monomer, possibly combined with problems during polymerization and fabrication. There is hope that a better synthetic route will lead to the commercialization of this interesting polymer [14, 15]. More recently, 3M introduced a novel approach to prepare perfluorinated ionomers using a starting material made by electrochemical fluorination (the "3M polymer") [16].

Several Japanese companies have also made advances in the design of electrolytic cells suitable for membrane technology. As a result of these improvements, membrane technology was widely adopted in Japan resulting in Japan having the most modern chlor-alkali industry in the world. While environmental considerations initially drove this development, further optimization of membrane construction and cell design soon demonstrated the superiority of the new technology in strictly economic terms: both capital and operating costs are lower and, in particular, electric power consumption is substantially reduced. Today, almost all new chlor-alkali installations use membrane technology. It is expected that eventually all older plants, now using obsolete technologies, will be phased out. This transition will not only eliminate the direct pollution inherent in the use of asbestos or mercury, but will also reduce the air pollution associated with the use of fossil fuels for the generation of electricity.

After the successful completion of the manned moon program in 1969, the interest in fuel cells subsided. Although the chlor-alkali application captured the limelight then, work on membrane development for fuel cells continued in the background. One such development was the introduction of liquid compositions of Nafion® [17, 18], which allowed casting of films much thinner than previously possible by extrusion. More importantly, it also led to the development of catalytic inks (platinum on carbon black dispersed in liquid Nafion®) by the Los Alamos National Laboratory. These inks have improved the utilization of platinum by a factor of more than 10. Without this improved platinum utilization, the price of platinum would have been prohibitive for use in polymer-based fuel cells. In 1995, Ballard Power Systems introduced a partially fluorinated polymer based on sulfonated polytrifluoro styrene [19]. This polymer is soluble in solvents such as dimethyl formamide or dimethyl sulfoxide.

Today there is a renewed interest in fuel cells. If the use of fuel cells for automotive applications materializes, then this application may again overshadow the chlor-alkali application. While there are many hurdles that must be overcome, the potential market is enormous: about 5 m^2 of ionomer area is needed for a 50-kW power plant. If this is multiplied by 100,000 vehicles, the area of ionomer required exceeds that for all other current applications combined.

The development of fluorinated ionomers began at the DuPont Experimental Station located on the banks of the Brandywine River, the very same location where the DuPont Company was started more than 200 years ago. Today Nafion® is made at a plant near Fayetteville, North Carolina. This plant uses the best available technology and has sufficient capacity to meet the foreseeable needs of both the chlor-alkali and the fuel cell markets. This is in keeping with the spirit of the founder of the DuPont Company, Eluther Irene DuPont, who said [20]:

> In constructing on the Brandywine, near Wilmington, Delaware, a manufacture of powder, I have wanted to make the establishment in every way worthy of the scale on which I have built it and I have tried to secure for it the best of all processes used in Europe, and to give my own attention to improving the refining of saltpetre as well as to any other changes that may effect the quality of the powder. My efforts have had some success and the reputation that my powder has already acquired is ample reward for the pains I have taken.

Figure 2.1 Photograph of an original DuPont powder mill (courtesy Hagley Museum and Library, Wilmington, Delaware).

References

1. Connolly, D.J., Gresham, W.F., US Patent 3,282,875 DuPont Co., Nov. 1, 1966.
2. Ebnesajjad, Sina, Melt processible fluoro polymers, Plastics Design Library, Norwich, NY, 2003.
3. Grot, W.G., US Patent 3,770,567 assigned to DuPont Co., Nov. 6, 1973.
4. Grot, W.G., US Patent 3,784,399 assigned to DuPont Co., Jan. 8, 1974.
5. Grot, W.G., US Patent 3,902,946 assigned to DuPont Co., Sept. 2, 1975.
6. Walmsley, P., US Patent 3,909,378 assigned to DuPont, Sep. 1975.
7. Grot, W.G., US Patent 4,026,783 assigned to DuPont Co., May 31, 1977.
8. Beer, H.B., GB Patent 1,147,442, 1965.
9. Bianchi, G., de Nora, V., Gallone, P., Nidola, A., US Patent 3,616,445 assigned to Electronor Corp., Oct. 26, 1971.
10. Oda, Y., Suhara, M., Endo, E., US Patent 4,065,366 assigned to Asahi Glass, Dec. 27, 1977.
11. Grot, W.G., Molnar, C.J., Resnick, P.R., US Patent 4,267,364 assigned to DuPont Co., May 12, 1981.
12. Carl, W.P., Ezzell, B.R., Mod, W.A., US Patent 4,358,412 assigned to Dow Chemical, Nov. 9, 1982.
13. Ghielmi, A. et al., Paper presented at the Grove Fuel Cell Symposium, Munich, Oct. 6, 2004.

14. Ghielmi, A., Vaccarono, P., Troglia, C., Arcella, V., J. Power Sources, **145**, 108–115, 2005.
15. Arcella, V., Ghielmi, A., Tommasi, G., (a) US Patent 6,767,977 assigned to Ausimont, July 27, 2004; (b) US Patent 6,639,011 assigned to Ausimont, Oct. 28, 2003.
16. Guerra, M.A., US Patent 6,624,328 assigned to 3M Innovative Properties Co., Sept. 23, 2003.
17. Grot, W.G., US Patent 4,433,082 assigned to DuPont Co., Feb. 21, 1984.
18. Grot, W.G., US Patent 4,453,991 assigned to DuPont Co., June 12, 1984.
19. Steck, A.E., Stone, C., Wei, J., US Patent 5,422,411 assigned to Ballard Power Systems, June 6, 1995.
20. DuPont, E.I., Letter to Secretary of State James Madison, 1804.

3 Manufacture

3.1 Introduction

Fluorinated ionomers are produced in relatively small volumes. For the economics of manufacture it is therefore important to integrate the manufacture with that of similar, larger volume products. For perfluorinated ionomers, at least the monomer synthesis can be integrated with that of other vinyl ether monomers such as perfluoro propyl vinyl ether (PPVE) or perfluoro methyl vinyl ether (PMVE). For partially fluorinated ionomers, there is the possibility of radiation grafting certain monomers on commercially available fluoropolymer films such as fluorinated ethylene propylene (FEP) copolymer. While this approach may lead to attractive economics, the oxidative stability of the products obtained has been poor.

3.2 Perfluorinated Ionomers

Perfluorinated ionomers are derived from melt-processable precursor polymers that are obtained by the copolymerization of tetrafluoroethylene (TFE) and a perfluorinated vinyl ether containing a sulfonyl halide or carboxylic ester functional group. The ether linkage is required to give the vinyl group sufficient activity to allow the incorporation of a high percentage of the functional monomers. The alkyl groups $-R_F$ used in the vinyl ethers $CF_2=CF-O-R_F$ are shown in Table 3.1. The molecular weight (MW) of the vinyl ether monomers is given and it is used as a short form to identify and distinguish them.

There are three large-scale, commercial producers of perfluorinated ionomers, namely DuPont, Asahi Glass and Asahi Chemicals (now named Asahi Kasei). All three of these companies now produce the same sulfonic polymer originally developed by DuPont. Asahi Kasei initially used a sulfonic ionomer containing one additional CF_2 group. A surface layer of carboxylic ionomer could be created on this polymer by a chemical posttreatment. Apparently, this approach has been abandoned now. Each of these companies also produces a carboxylic ionomer, because both sulfonic and carboxylic ionomers are needed for chlor-alkali membranes,

Table 3.1 Alkyl Groups Used in Perfluorinated Vinyl Ethers

MW	$-R_F$	Polymer Name	Company

446	$-CF_2-CF-O-CF_2-CF_2-SO_2F$
	       $|$
	     CF_3
		Nafion®
		 Flemion®
		 Aciplex (now)
			DuPont
			 Asahi Glass
			 Asahi Chemicals

496	$-CF_2-CF-O-CF_2-CF_2-CF_2-SO_2F$
	      $|$
	     CF_3
		Aciplex (old)	Asahi Chemicals

422	$-CF_2-CF-O-CF_2-CF_2-CO_2-CH_3$
	      $|$
	     CF_3
		Nafion®
		 Aciplex
			DuPont
			 Asahi Chemical

306	$-CF_2-CF_2-CF_2-CO_2-CH_3$	Flemion®	Asahi Glass
380	$-CF_2-CF_2-CF_2-CF_2-SO_2F$	3M polymer	3M
280	$-CF_2-CF_2-SO_2F$	Dow polymer Hyflon-Ion	Dow Solvay-Solexis

currently by far the most important application of perfluorinated ionomers. The steps involved in the large scale manufacture of perfluorinated ionomer products, such as NAFION®, FLEMION® and ACIPLEX®, are shown in Fig. 3.1 An experimental perfluorinated ionomer (MW = 280), first introduced by Dow Chemical, is now offered by Solvay-Solexis under the name Hyflon-Ion [12]. More recently, Minnesota Mining and Manufacturing (3M) published the synthesis of another sulfonic monomer containing only a single ether linkage [14].

3.2.1 Monomer Synthesis

The important raw materials for the preparation of perfluorinated ionomers are TFE and hexafluoro propylene epoxide (HFPO). Their synthesis is discussed in brief.

The synthesis of TFE starts with the reaction of chloroform (obtained by the chlorination of methane) with anhydrous hydrofluoric acid (HF) to yield chlorodifluoromethane. Pyrolysis of this compound yields TFE and a by-product (HCl) [1]. TFE can undergo autopolymerization if it is not inhibited. Effective inhibitors of TFE autopolymerization include a variety

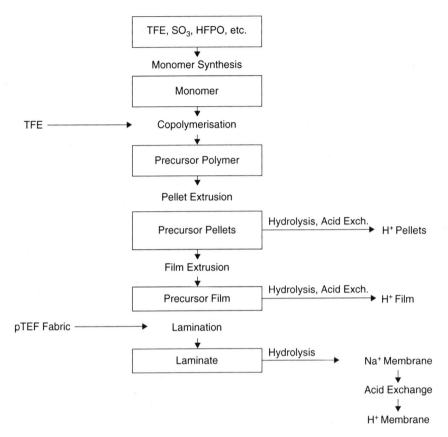

Figure 3.1 Steps in the manufacture of various ionomer products.

of terpenes such as α-pinene, terpene B, and D-limonene, which appear to act as scavengers of oxygen, a polymerization initiator.

3.2.1.1 Properties of TFE

TFE is highly flammable and can undergo violent deflagration even in the absence of air:

$$C_2F_4 \rightarrow C + CF_4$$

Heat of reaction values between 57 and 62 kcal/mol (at 25°C and 1 atm) have been reported for TFE deflagration. For shipment, TFE can be stabilized by diluting it with carbon dioxide or anhydrous HCl.

Table 3.2 lists the properties of TFE. It is a colorless, odorless, taste-less, non-toxic gas which boils at −76.3°C and freezes at −142.5°C. Its critical temperature and pressure are 33.3°C and 39.2 MPa, respectively. TFE is stored as a liquid and its vapor pressure at −20°C is 1 MPa. Its heat of formation is reported to be −151.9 kcal/mol. Polymerization of TFE is highly exothermic and generates 41.12 kcal/mol of heat. The extent of exothermicity of TFE polymerization is evident when it is compared with the polymerization of vinyl chloride and styrene, which have heats of polymerization of 23–26 and 16.7 kcal/mol, respectively.

A complete description of the hazards of TFE can be found in Ref. [1]. Safe storage of TFE requires its oxygen content to be less than 20 ppm. The temperature and pressure should be controlled during its storage.

Table 3.2 Properties of TFE

Property	Value
Molecular weight	100.02
Boiling point at 101.3 kPa, °C	−76.3
Freezing point, °C	−142.5
Liquid density vs. temperature (°C), g/mL	
$-100 < t-40$	$1.202 - 0.0041t$
$-40 < t < 8$	$1.1507 - 0.0069t - 0.000037t2$
$8 < t < 30$	$1.1325 - 0.0029t - 0.00025t2$
Vapor pressure at T K, kPa	
$196.85 < T < 273.15$	$\log_{10}P_{kPa} = 6.4593 - 875.14/T$
$273.15 < T < 306.45$	$\log_{10}P_{kPa} = 6.4289 - 866.84/T$
Critical temperature, °C	33.3
Critical pressure, MPa	3.92
Critical density, g/mL	0.58
Dielectric constant at 28°C	
at 101.3 kPa	1.0017
at 858 kPa	1.015
Thermal conductivity at 30°C, mW/(m·K)	15.5
Heat of formation for ideal gas at 25°C, ΔH, kJ/mol	−635.5
Heat of polymerization to solid polymer at 25°C, ΔH, kJ/mol	−172.0
Flammability limits in air at 101.3 kPa, vol.%	14–43

Increasing the temperature, especially at high pressures, can initiate deflagration in the absence of air. TFE forms explosive mixtures with oxygen or air. Detonation of a mixture of TFE and oxygen can increase the maximum pressure to 100 times the initial pressure.

Hexafluoropropylene (HFP, b.p. −29°C) is obtained by the pyrolysis of TFE at 700–800°C under reduced pressure [2] or dilution with carbon dioxide [3], particularly in the presence of 3–10% chlorodifluoromethane [4].

HFPO (b.p. −27°C) is a key ingredient in the synthesis of most of the functional monomers used. It can be prepared by oxidation of HFP using alkaline hydrogen peroxide below 0°C [5] or with oxygen at 100–200°C [6]. Hoechst has patented a process for making HFPO by the anodic oxidation of HFP. This process uses a cylindrical cell with a lead dioxide anode and a Nafion separator. It is described in more detail in Section 5.1.13.

The synthetic routes to all except the last monomer (MW = 280) in Table 3.1 utilize the same last steps to introduce the perfluoro vinyl ether function: addition of HFPO to a perfluoro acyl fluoride followed by dehalocarbonylation. This is the same chemistry that is used in the synthesis of other perfluorinated vinyl ether monomers such as PMVE and PPVE. The synthesis of the 3M monomer (MW = 380) is used to illustrate this reaction sequence; other monomers are prepared by using different perfluoro acyl fluorides:

$$
\underset{\underset{\text{CF}_3}{|}}{\text{FSO}_2\text{CF}_2\text{CF}_2\text{CF}_2\text{COF} + \text{CF}-\text{CF}_2} \xrightarrow{\text{KF}} \underset{\underset{\text{Na}_2\text{CO}_3 \big\downarrow \quad \text{CF}_3}{|}}{\text{FSO}_2\text{CF}_2\text{CF}_2\text{CF}_2\text{CF}_2\text{OCFCOF}}
$$

$$
\text{FSO}_2\text{CF}_2\text{CF}_2\text{CFCF}_2\text{OCF}=\text{CF}_2 + 2\text{NaF} + 2\text{CO}_2
$$

If the dehalocarbonylation is applied to a sulfonyl fluoride intermediate of suitable chain length, an undesired side reaction involving the abstraction of the fluoride from the −SO$_2$F group will occur, resulting in the formation of a five- or six-membered ring (cyclic sulfone). To avoid this side reaction, the first synthesis described (MW = 446) uses the addition of 2 mol of HFPO and the 3M monomer (MW = 380) starts with butane sultone instead of propane sultone. In the synthesis of the Dow monomer (MW = 280), the problem is avoided by using the more reactive chloro difluoro methyl group in the α-position.

3.2.1.2 Perfluoro 3,6-dioxa-4-methyl-7-octene sulfonyl fluoride (MW = 446)

This monomer is used in the manufacture of Nafion, Flemion and Aciplex sulfonic acid polymer (see Table 3.1). Among the six monomers shown in Table 3.1, it is produced on the largest scale. In addition, it has received the widest attention from many research groups because Nafion has been readily available (for instance from Aldrich Chemicals) for many years.

Safety

The synthesis of this monomer involves extreme hazards. The following summary should serve only as a preliminary guide and not as a substitute for a thorough safety review of the process. The flammability and explosivity of TFE have been mentioned above. Sulfur trioxide (b.p. 45°C) reacts explosively with water and many other compounds. It is extremely destructive to human tissue, clothing, etc., and emits choking fumes.

In the first step of the synthesis, sulfur trioxide not only adds to the TFE double bond but it also can act as an oxidant in a very exothermic reaction yielding carbonyl fluoride (more toxic than phosgene) and sulfur dioxide. Mixtures of TFE and sulfur trioxide are therefore explosive. While the solubility, and therefore the concentration, of TFE in the reaction mixture under atmospheric conditions is small, mixtures of sulfur trioxide with the cyclic addition product are also explosive for the same reason:

$$\mathrm{SO_3} + \begin{array}{c} \mathrm{CF_2-CF_2} \\ | \qquad | \\ \mathrm{SO_2-O} \end{array} \longrightarrow 2\,\mathrm{SO_2} + 2\,\mathrm{COF_2}$$

At about 50% conversion, the reaction mixture reaches maximum explosivity.

The cyclic sultone and the acid fluorides formed in subsequent reactions react with water releasing HF and the sultone violently, the higher acid fluorides more slowly,, resulting in the formation of highly toxic perfluorinated acids.

3.2.1.3 TFE Sultone (3,3,4,4-tetrafluoro-1,2-oxathietane S,S-dioxide)

Sulfur trioxide exists in several molecular forms. Of these only the monomeric form will give the desired product in this reaction. Therefore, it is essential that this material is freshly distilled immediately before the reaction with TFE. Traces of moisture will catalyze the polymerization of sulfur trioxide. From a safety standpoint, it is best to use single apparatus for the entire operation, including the distillations of the sulfur trioxide and the TFE sultone formed. In this way, the entire operation can be remotely controlled by turning on the heating mantels, and cooling water and TFE gas.

The reaction is conducted at room temperature under atmospheric conditions. As discussed above, sulfur trioxide is distilled directly into the reaction vessel. Sparging of TFE gas can begin immediately when the sparger is covered with liquid, even before the distillation process of sulfur trioxide is complete. A column filled with silica gel can be used to remove D-limonene inhibitor from the TFE. The rate of reaction is limited by the low solubility of TFE in the reaction medium. As the reaction is slightly exothermic and an increase in temperature will further reduce TFE solubility, cooling to maintain a temperature of 20–30°C will speed up the reaction.

HFPO addition

The addition of HFPO to a perfluorinated acid fluoride is an important step in many synthetic routes to perfluorinated ionomers; hence, it will be discussed in some detail, using the Nafion monomer as an example.

The reaction is catalyzed by fluoride ions; CsF is preferred because of its solubility in solvents such as diglyme (diethylene glycol dimethyl ether), tetraglyme, acetonitrile and adiponitrile. TFE sultone can be used for the reaction directly, or after fluoride ion catalyzed ring opening:

$$\begin{array}{c} CF_2-CF_2 \\ | \quad \ \ | \\ SO_2-O \end{array} + F^- \longrightarrow FSO_2CF_2CF_2O^- \ \rightleftharpoons \ FSO_2CF_2COF + F^-$$
$$bp. = 30°C$$

The equilibrium between a perfluorinated alkoxide anion and the corresponding acyl fluoride, indicated by the double arrow, is an important factor for the subsequent addition of HFPO. This equilibrium is in general far on the side of the acyl fluoride (the right-hand side of the equation); however, a negative substitution – such as the sulfonyl fluoride group – favors the alkoxide ion relative to other acyl fluorides that lack this substitution. As this alkoxide anion attaches to the central carbon of the HFPO molecule, the first step in HFPO addition occurs with little competition:

$$
\begin{array}{c}
\text{O} \\
\diagup\ \diagdown \\
FSO_2CF_2CF_2O^- + CF{-}CF_2 \longrightarrow FSO_2CF_2CF_2OCF{-}CFO^- \text{(mono adduct)} \\
| \qquad\qquad\qquad\qquad | \\
CF_3 \qquad\qquad\qquad\qquad\ CF_3
\end{array}
$$

After all of the alkoxide based on the TFE sultone has been used up, the reaction proceeds less selectively because all the acyl fluorides involved have essentially the same affinity to fluoride ion. Therefore, the desired addition of a second molecule of HFPO has to compete with the undesired addition of a third molecule of HFPO and with the oligomerization of HFPO itself; it is also catalyzed by fluoride ions according to the same mechanism. To obtain a high yield of di-adduct it is therefore desirable to stop the reaction at relatively low conversion of mono- to di-adduct and to distill the reaction mixture to recover both mono- and di-adducts. The mono-adduct is recycled back to the HFPO addition:

$$
\begin{array}{c}
\text{O} \\
\diagup\ \diagdown \\
FSO_2CF_2CF_2OCF{-}CF_2O^- + CF{-}CF_2 \longrightarrow FSO_2CF_2CF_2OCF{-}CF_2OCFCF_2O^- \text{(diadduct)} \\
| \qquad\qquad | \qquad\qquad\qquad\qquad\quad | \qquad\quad | \\
CF_3 \qquad\quad CF_3 \qquad\qquad\qquad\qquad CF_3 \quad\ CF_3 \quad \text{b.p. = 140°C}
\end{array}
$$

After the addition of slightly more than 1 mol of HFPO, phase separation occurs with the lower phase containing more higher MW fluorocarbons and less diglyme and catalyst. The upper phase contains more diglyme, catalyst and mono-adduct. The di-adduct is therefore somewhat protected against further HFPO addition. At the end of the reaction, it is preferred to recover only the lower layer for distillation; a stopcock at the bottom of the reactor is convenient for this purpose. The upper layer is retained for the next batch. A review of the HFPO chemistry can be found in Ref. [8].

The final step in the monomer synthesis is the reaction of the di-adduct with an oxygen-containing sodium or potassium salt of a weak acid. Any salt of a weak acid that can be expressed by an anhydride formula, such as

Na_2O-CO_2 for sodium carbonate, appears to be suitable. Sodium carbonate is commonly used either as a heated, packed bed of pellets through which vapors of di-adduct are passed at about 300°C or as a slurry in boiling diglyme. Sodium silicate ($Na_2O-nSiO_2$) in the form of glass beads has the advantage that it can be used as a fluidized bed:

$$FSO_2CF_2CF_2OCFCF_2OCFCF=O + Na_2CO_3 \longrightarrow FSO_2CF_2CF_2OCFCF_2OCF=CF_2 + 2NaF + 2CO_2$$

with CF_3 and CF_3 substituents on the left reactant and CF_3 on the product.

If the reaction is attempted on the mono-adduct, a cyclic compound is formed instead.

Example from Ref. [7]

Additional examples of HFPO additions to an acyl fluoride can be found in Ref. [11].

In a dry flask, 2.9 g of dry cesium fluoride is placed. The flask was attached to a manifold, cooled in an ice bath and evacuated; 40 ml of dry diglyme and 50 g of TFE sultone were injected into the flask through a side arm. Then 115 g of HFPO was pressured into the flask through a reduction valve set at 4 psig. The uptake of the entire amount of HFPO took 30 min. The lower fluorocarbon layer (158 g) was separated and distilled; 40 g of a fraction boiling at 138–141°C (fraction A) and 38 g of a fraction boiling at 179–181°C were obtained. IR and NMR spectra as well as elemental analysis have identified fraction A as the adduct of 2 mol of HFPO with one of TFE sultone with the following structure:

$$FSO_2CF_2CF_2OCF-CF_2OCFCOF$$

with CF_3 and CF_3 substituents.

Fraction B was identified as the corresponding adduct of 3 mol of HFPO.

3.2.1.4 *Methyl perfluoro 5-oxa heptenoate (MW = 306)*

1,4-Diodo octafluoro butane (b.p. 85°C at 100 mmHg, obtained by the telomerization of TFE with iodine) is added to fuming sulfuric acid (15–30% sulfur trioxide) at 90°C. Perfluoro gamma butyrolactone (b.p. 18°C) is formed [9].

In the next step, the lactone is caused to react with an alcohol, such as methanol or ethanol, to form a perfluorinated acyl fluoride containing an ester group. The reaction is highly exothermic. An excess of alcohol would lead to the formation of a diester. The lactone is therefore charged to the reactor first and then the alcohol is added gradually with vigorous stirring to avoid a localized excess. Low temperature (–40°C to +20°C) and a diluent (such as diglyme) are used for both the lactone and the alcohol to avoid localized formation of diester. The reaction results in the formation of HF, and sodium fluoride (NaF) is used as a HF scavenger. Anhydrous conditions should be maintained to avoid the formation of carboxylic acid groups. Glass equipment may be etched by the by-product HF; stainless steel or Hastelloy equipment is preferred. The boiling points for the products are 96°C and 111°C for the methyl and ethyl esters, respectively.

Example 1 from Ref. [10]

In an autoclave made of stainless steel, 208 g (4.95 mol) of NaF (pure) was charged and dried. In the autoclave, 400 ml of anhydrous diethylene-glycol dimethyl ether (diglyme) was charged and the mixture was cooled to –40°C; and then 400 g (2.26 mol) of perfluoro-.gamma.-butyrolactone was added in a liquid form.

The temperature in the autoclave was maintained at –40°C, and a mixture of ethanol and diethyleneglycol dimethyl ether (2.5/1.0 by volume of the solvent/ethanol) was continuously fed with vigorous stirring until providing a 1.2 M ratio of ethanol to perfluoro-.gamma.-butyrolactone. The reaction was ceased at this ratio.

The reaction mixture in the autoclave was filtered to separate the solid from the liquid, and the filtrate was distilled to obtain 3-carboethoxy per-fluoropropionyl fluoride having a b.p. of 110–112°C (at 760 mmHg) in the yield of 64 mol. % (based on perfluoro-.gamma.-butyrolactone).

The next two steps, HFPO addition and dehalocarbonylation, are similar to the corresponding steps in the synthesis of the first monomer (MW = 446) described, except that the dehalocarbonylation is carried out on the dried alkali metal salt obtained from the HFPO adduct. The following examples are taken from Ref. [11].

Comparative Example 1

Preparation of 6-carboethoxy-perfluoro-2-methyl-3-oxa-hexanoyl fluoride

In a 1-1 autoclave equipped with a stirrer, 17.7 g (0.116 mol) of anhydrous cesium fluoride powder, 188 g of tetraethyleneglycol dimethyl ether and 240 g (1.09 mol) of 3-carboethoxy-perfluoropropionyl fluoride having the formula:

$$O=CCF_2CF_2CO_2-CH_2CH_3$$
$$|$$
$$F$$

were charged. The mixture was vigorously stirred at a reaction temperature of −10°C to 0°C under a pressure of less than 1 kg/cm^2, and 211 g (1.27 mol) of hexafluoropropylene oxide was added continuously for 3 h.

After the reaction, the reaction mixture was distilled to obtain 240 g of the object compound of 6-carboethoxy-perfluoro-2-methyl-3-oxa-hexanoyl fluoride (b.p.: about 53°C at 15 mmHg; yield: 57%) and to recover 44 g of the starting material of 3-carboethoxy-perfluoropropionyl fluoride.

Comparative Example 2

Preparation of 9-carboethoxy-perfluoro-2,5-dimethyl-3, 6-dioxa-nonanoyl fluoride

In a 1-1 autoclave equipped with a stirrer, 7.6 g (0.05 mol) of anhydrous cesium fluoride powder, 42 g of tetraethyleneglycol dimethyl ether and 220 g (1.0 mol) of 3-carboethoxy-perfluoropropionyl fluoride having the formula

$$O=CCF_2CF_2CO_2-CH_2CH_3$$
$$|$$
$$F$$

were charged. The mixture was vigorously stirred at a reaction temperature of −10°C to 0°C under a pressure of about 1 kg/cm^2, 380 g (2.2 mol) of hexafluoropropylene oxide was added continuously for 6 h.

After the reaction, the reaction mixture was distilled to obtain 309 g of the object compound of 9-carboethoxy-perfluoro-2,5-dimethyl-3,6-dioxa-nonanoyl fluoride (b.p.: about 66°C at 8 mmHg; yield: 56%).

Comparative Example 3

Preparation of 6-carbomethoxy-perfluoro-2-methyl-3-oxa-hexanoyl fluoride

In a 1-l autoclave equipped with a stirrer, 15.2 g (0.1 mol) of anhydrous cesium fluoride powder, 70 ml of diethyleneglycol dimethyl ether, 206 g (1.0 mol) of 3-carbomethoxy-perfluoropropionyl fluoride having the formula:

$$O{=}CCF_2CF_2CO_2{-}CH_3$$
$$|$$
$$F$$

were charged. The mixture was vigorously stirred at a reaction temperature of –20°C to –10°C under a pressure of less than 1 kg/cm^2, and 186 g (1.12 mol) of hexafluoropropylene oxide was added continuously for 4 h. After the reaction, the reaction mixture was distilled to obtain 208 g of the object compound of 6-carbomethoxy-perfluoro-2-methyl-3-oxa-hexanoyl fluoride (b.p.: 68°C at 65 mmHg; yield: 56%) and to recover 39 g of 3-carbomethoxy-perfluoropropionyl fluoride.

Example 2

In a 100-ml four-necked flask equipped with a reflux condenser, a dropping funnel and a magnetic stirrer, which was purged with nitrogen gas, 8.24 g of sodium carbonate dried at 280°C for 2 h and 40 ml of anhydrous diethyleneglycol dimethyl ether were charged.

The mixture was stirred and 20.0 g of 6-carboethoxy-perfluoro-2-methyl-3-oxa-hexanoyl fluoride of Comparative Example 1 was added dropwise at room temperature for 3 h.

After the addition, the stirring was continued for further 1 h and the solvent of diethyleneglycol dimethyl ether was distilled off. The residue was dried at 80°C at 3 mmHg for 2 h to obtain 25.2 g of a solid mixture of the object compound having the formula:

$$NaCO_2CFOCF_2CF_2CF_2CO_2{-}CH_2CH_3$$
$$|$$
$$CF_3$$

and NaF and sodium carbonate.
According to the analysis, the amount of the object compound was 20.3 g and the yield was 97%.

Example 3

In a 100 ml round-bottomed flask, 21 g of the solid object compound of Example 1 was charged. The flask was connected to a series of traps: one trap maintained at –78°C and another at –196°C, followed by a vacuum pump. The reaction was carried out at a reaction temperature of 170°C under a reduced pressure of 3 mmHg for 3 h with stirring.

In the trap maintained at –78°C, 14.9 g of a liquid was collected. The liquid was carefully distilled, and 10.1 g of ethyl perfluoro-5-oxa-6-heptenoate was obtained:

$$CF_2 = CFOCF_2CF_2CF_2CO_2-CH_2CH_3$$

Example 4

In the apparatus of Example 3, 23.0 g of the solid mixture of the object compound, potassium fluoride and potassium carbonate, which was obtained by the reaction of 17.0 g of the starting material and 8.3 g of potassium carbonate, was charged, and the reaction was carried out at 150–155°C under a reduced pressure of 2–4 mmHg for 3 h with stirring.

In the trap maintained at –78°C, 13.3 g of a liquid was collected.

The liquid was distilled in accordance with the process of Example 3 and 9.2 g of ethyl perfluoro-5-oxa-6-heptenoate was obtained.

The chemistry of this monomer, its copolymerization with TFE and the properties of the resulting polymer are the subject of an excellent review [13].

3.2.1.5 Perfluoro-4-(fluorosulfonyl)butylvinyl ether (MW = 380)

The following procedure is taken from US Patent 6,624,328 [14]. While the patent also claims propane sultone as starting material, its use results in poor yields in the final step due to the formation of a cyclic sultone:

1,4-Butane sultone (6900 g, 50.7 m) was electrochemically fluorinated in HF as described in US Patent 2,732,398 to obtain 4-(fluorosulfonyl) hexafluorobutyryl fluoride, $FSO_2CF_2CF_2CF_2COF$ (4000 g, 14.3 m for a 28% yield).

In the next step, this acyl fluoride was caused to react with HFPO according to the following equation:

$$FSO_2CF_2CF_2CF_2COF + \overset{\displaystyle O}{\overset{\displaystyle /\ \backslash}{\underset{\displaystyle |}{\underset{\displaystyle CF_3}{CF-CF_2}}}} \xrightarrow{\ KF\ } FSO_2CF_2CF_2CF_2CF_2OCFCOF$$
$$\underset{\displaystyle CF_3}{|}$$

2162 g (7.7 m) of 4-(fluorosulfonyl)hexafluorobutyryl fluoride, $FSO_2CF_2CF_2CF_2COF$ was reacted with an equimolar amount of hexafluoropropylene oxide (HFPO) (1281 g, 7.7 m) in 2 l diglyme with 114 g potassium fluoride to obtain perfluoro-4-(fluorosulfonyl)butoxypropionyl fluoride (2250 g, 5.1 m for a 65% yield) and 675 g of a higher boiling by-product that had an additional hexafluoropropylene oxide unit.

This is followed by dehalocarbonylation as shown in the following equation:

$$FSO_2CF_2CF_2CF_2CF_2OCFCOF \ + Na_2CO_3 \rightarrow FSO_2CF_2CF_2CF_2CF_2OCF{=}CF_2 + 2NaF + 2CO_2$$
$$\underset{CF_3}{|}$$

1108 g (2.5 m) of perfluoro-4-(fluorosulfonyl)butoxypropionyl fluoride was then reacted with sodium carbonate (603 g, 5.7 m) in glyme at 70°C to make the sodium salt of the acid. Solvent was then removed under vacuum and the dried salt was heated to 165°C to break vacuum with the carbon dioxide by-product and continuing to heat up to 182°C to isolate perfluoro-4-(fluorosulfonyl)butylvinyl ether (703 g, 1.9 m for a 74% yield).

3.2.1.6   Perfluoro-3-oxa-4-pentene sulfonyl fluoride (MW = 280)

The original Dow synthetic route was based on the observation that adducts derived from the epoxide of perfluoro allylchloride (instead of HFPO), on dehalocarbonylation, result in less ring closure and therefore allow the formation of this low MW monomer in reasonable yield [15]. The same inventors describe the addition of the epoxide of perfluoro allylchloride to acid fluorides in Ref. [16]. There are two known routes to the required perfluoro allylchloride: the zinc dehalogenation of perfluoro 1,2,3-trichloro propane (b.p. = 72°C, from Halocarbon Products) and the pyrolysis of a mixture of chloro difluoro methane and chloro trifluoro ethylene. The zinc dehalogenation is conducted at about 70°C in iso-propanol as a solvent and gives a good yield (about 95%). However, the disposal of by-product zinc chloride represents a serious problem. The subsequent epoxidation, conducted with molecular oxygen at 110°C in perfluoro trichloro propane as a solvent, yields only about 50% of the desired epoxide; the balance is mostly difluoro chloro acetyl fluoride and some carbonyl fluoride. The yield in the subsequent dehalocarbonylation is 67%, with the balance consisting mostly of the cyclic sulfone. The b.p. of the monomer is 77°C.

Solvay-Solexis has developed an easier route to the same monomer. The starting product is again the cyclic TFE sultone. Instead of using a catalytic amount of fluoride ions for opening the four-membered ring, a stoichiometric amount of elemental fluorine is used, resulting in the formation of a perfluorinated hypofluorite:

$$CF_2 \quad CF_2$$
$$| \quad \quad | \quad \quad + F_2 \quad \longrightarrow \quad FSO_2CF_2CF_2OF$$
$$SO_2-O$$

This hypofluorite is added across the double bond of 1,2-dichloro difluoro ethylene, followed by dehalogenation to yield the desired vinyl ether:

$$FSO_2CF_2CF_2OF \quad + \quad ClCF=CFCl \longrightarrow FSO_2CF_2CF_2OClCF-CF_2Cl$$

$$FSO_2CF_2CF_2OClCF-CF_2Cl \rightarrow FSO_2CF_2CF_2OCF=CF_2$$

3.2.1.7 *Perfluoro-4,7-dioxa-5-methyl-8-nonene sulfonyl fluoride (MW = 496)*

This monomer appears to be no longer in use. The following procedure is taken from US patent 4,597,913 [17]:

Example 1: TFE is reacted with sodium ethyl mercaptide and dimethyl carbonate dissolved in tetrahydrofuran to yield: $C_2H_5-S-CF_2-CF_2-CO_2CH_3$. At this point, two options are offered:
Example 2: Hydrolysis of the methyl ester groups followed by reaction with sulfur tetrafluoride to yield acyl fluoride $C_2H_5-SCF_2CF_2-COF$ or:
Example 3: Oxidation of the thioether linkage to the sulfone using hydrogen peroxide. This is followed by
Example 4: Conversion to an acyl fluoride similar to Example 2. HFPO is then added to either acyl fluoride:
Examples 11 and 13, respectively, followed by dehalocarbonylation using sodium carbonate (Examples 12 and 16). The resulting vinyl ethers are copolymerized with TFE, fabricated to a film and the thioether linkage converted to a sulfonyl chloride group using chlorine gas at 120°C.

3.2.1.8 *Methyl perfluoro-3,6-dioxa-4-methyl noneoate (MW = 422)*

Methyl 3-methoxy tetrafluoro propionate is prepared by the addition of dimethyl carbonate to TFE using sodium methoxide as a catalyst; 100 g of this material is added dropwise to 40 ml of sulfur trioxide at a rate to maintain a gentle reflux due to the exothermic reaction. The reaction mixture was distilled at atmospheric pressure and the fraction boiling from 82°C to 86°C was passed over NaF pellets at 400°C and 4 mmHg pressure to convert by-product methyl fluorosulfonate to methyl fluoride and sodium sulfate. Pure methyl difluoro malonyl fluoride was then isolated by distillation.

In the next step, 29.2 g HFPO was slowly added to 26.3 g methyl difluoro malonyl fluoride dissolved in 20 ml of tetraglyme containing 6 g CsF as catalyst. The mono HFPO adduct was isolated by distillation: b.p. 75–95°C at 400 mmHg.

The dehalocarbonylation was carried out in a vertical stirred bed reactor filled with sodium phosphate at 235–240°C with a slow stream of nitrogen. The product formed was isolated by distillation: b.p. 98°C at 110 mmHg [18].

3.3 Polymerization

The copolymerization can be carried out in an aqueous or non-aqueous medium. The aqueous polymerization can be carried out in a process similar to the one used to make aqueous poly tetrafluoro ethylene (pTFE) dispersions. A horizontal autoclave equipped with a slow-moving (about 100 rpm) paddle wheel stirrer can be used. A perfluorinated surfactant, such as C-8 (the ammonium salt of perfluoro octanoic acid) or the corresponding sulfonate salt (FC 95 made by 3M), is used. Ammonium persulfate can be used as an initiator. The comonomer forms a separate phase at the bottom of the autoclave throughout the run. TFE is fed to the reactor on demand at a certain pressure, and this TFE pressure determines the EW. However, a lower EW can be achieved if part of the comonomer is predispersed in the surfactant solution by a high-speed mixer. The polymerization is stopped when the desired amount of TFE has been consumed. The lower layer of unreacted comonomer can then be recovered and the aqueous dispersion of the ionomer precursor coagulated.

With more vigorous agitation, the bottom layer of comonomer is completely dispersed in the aqueous phase. This system is described in detail by Quarderer and Katti [19]. Again, gaseous TFE is fed on demand to the polymerization vessel containing all other components as a single, initial charge. The liquid phase in the vessel consists of a continuous aqueous phase containing dissolved initiator and a pH buffer. Most of the surfactant is present as micelles containing about 50 surfactant molecules. They serve as a reservoir of surfactant needed in the later stages of the polymerization. The comonomer is dispersed as droplets stabilized by a layer of surfactant on the outside. Three distinct stages of the polymerization are recognized: During the first stage, free radicals generated in the aqueous phase cause the formation of oligomer molecules, which enter either a dispersed monomer droplet or a micelle depending on the relative surface area of these two dispersed entities. As the size of these dispersed particles grows, the surfactant needed to cover their surface is obtained by the elimination of some micelles. The first stage is completed when all surfactant micelles are either eliminated or converted to growing polymer particles. The second stage is completed when all comonomer droplets are also converted to growing polymer particles.

The reference provides a thorough mathematical analysis of the system, but only limited connections to real polymerizations. Figures 5 and 6 of the reference show the EW of two unidentified precursor ionomers, MA and MB, for three different unspecified temperatures as a function of the unspecified TFE pressure. By extrapolating the curves to what is believed zero TFE pressure, one obtains an EW of about 330 for the MA homopolymer and 600 for MB homopolymer. From this, one may conclude that polymer MA is the Dow precursor and polymer MB is the Nafion precursor.

Ausimont has patented a "microemulsion" process, in which an emulsion of a perfluorinated polyether of an MW of about 500 is emulsified in water using a perfluorinated surfactant. After adjusting the pH of this emulsion to 4–7, it is added to the aqueous polymerization reaction together with the sulfonyl fluoride monomer and an initiator, such as a persulfate, and gaseous TFE on demand under a constant pressure [20]. Increased productivity is claimed. It is believed that Solvay-Solexis uses this process in the production of Hyflon-Ion.

Sulfate radicals formed by the cleavage of a persulfate ion initiate the aqueous polymerization. The sulfate radical attaches itself through an oxygen linkage to a TFE molecule, forming a new anion radical:

$$SO_4^- \bullet \ + \ TFE \ \longrightarrow \ ^-SO_3-O-CF_2-CF_2 \bullet$$

The chain propagation continues with the addition of TFE and vinyl ether monomers in a random fashion. The addition of a vinyl ether monomer can occur either at the carbon attached to the oxygen or at the terminal carbon. In the latter case, the resulting radical can undergo chain transfer:

$$-CF_2-CF_2\bullet \ + CF_2=CF-O-R_f \ \longrightarrow \ -CF_2-CF_2-CF_2-CF=O \ + R_f \bullet$$

The acyl fluoride formed will undergo hydrolysis in the aqueous system. Therefore, the sulfuric acid half ester will be formed in the initiation step. Carboxylic acid groups will then terminate the polymer chain formed by this mechanism at both the ends. These groups are unstable particularly in fuel cell applications. An iron–bisulfite initiator on the other hand will form a sulfonic acid group, which is stable, at the beginning of the polymer chain, thereby resulting in fewer unstable end-groups. In either case, end group stabilization, usually by fluorination, will be required for ionomer precursor polymers intended for fuel cell applications.

For the non-aqueous polymerization, perfluorinated solvents, such as perfluoro dimethyl cyclobutane (HFP dimer), or various perfluorinated liquids offered by 3M are preferred. F-113, a much less expensive solvent, can be used with a small loss in MW. However, at temperatures above 85°C, F-113 acts as a telomerizing agent. Perfluoro propionyl peroxide (3P) or N_2F_2 are suitable initiators. TFE is fed to the reactor on demand at a constant pressure. The TFE pressure and the concentration of comonomer determine the EW of the polymer formed. If the TFE pressure is changed to make an EW adjustment, then, under otherwise identical conditions, the meltflow of the polymer will also change (the meltflow will decrease and the EW will increase if the TFE pressure is raised). As the EW of the polymer formed depends on the comonomer concentration, conversion of comonomer has to be limited to prevent a broad distribution of the EW. Also, as the precursor polymers formed have only a limited solubility in the solvent/monomer mixture, the concentration of polymer formed should not exceed about 15% to maintain good agitation. When the desired concentration of polymer in the reaction slurry is achieved, the unreacted TFE is recovered by venting and the unreacted comonomer by evaporation. The non-aqueous copolymerization of the Nafion and Dow monomers with TFE in F-113, perfluoro dimethyl cyclobutane (HFP dimmer) or the liquid monomer as solvent, and 3P as the initiator, are described in Ref. [21].

3.4 Fabrication

The largest use for fluorinated ionomers is as unreinforced films or as fabric reinforced membranes. Extrusion of the polymer in the precursor form is the most commonly used method of fabrication. An alternative, namely the casting of films and membranes from solution, is discussed in Section 3.6.

Starting with polymer in the fluff form, it is common practice to first extrude a cylindrical shape of a few millimeters diameter, which is then chopped to a cylindrical pellet. This allows the removal of volatiles and converts the fluff into a shape that can be fed into an extruder at a controlled rate. The pellets are then fed into an extruder to produce the precursor film. Due to the high comonomer content of the precursor resin, the melting point of the resin is fairly low with a correspondingly low extrusion temperature, typically 280°C. The extruded precursor film, for the same reason, is soft and tacky, and requires an interleave film to prevent sticking to itself. Sulfonic/carboxylic bifilms can be obtained by coextrusion, where the lower-density carboxylate ester polymer floats on top of the heavier sulfonyl fluoride polymer. As the possible release of trace amounts of HF, Hastelloy is the preferred material of construction for the extruder. The extruded precursor film can be subjected to hydrolysis and acid exchange, or it can be first laminated to a fabric reinforcement.

Experimental quantities of polymer, too small to fill an extruder, can be converted to film by hot pressing or by skiving of the precursor form. While industrial skiving uses a cylindrical billet on a lathe, for laboratory purposes it is easier to mold a flat billet of about 5-mm thickness onto the roughened surface of a metal block. The skiving is then done on a shaper (a machine tool capable of performing reciprocating horizontal strokes over the surface of the work piece). Skiving is faster and results in better thickness uniformity than hot pressing.

The fabric reinforcement is woven from pTFE fibers or yarn to match the excellent chemical resistance of perfluorinated ionomers. Because pTFE is not meltfabricable, two special methods have been devised to convert this polymer into filaments: in the first, an aqueous dispersion of pTFE is added to a cellulose xanthate solution obtained by treating cellulose with sodium hydroxide and carbon disulfide. This mixture is then used in the standard viscose rayon spinning process to obtain fibers of regenerated cellulose containing imbedded pTFE particles. These fibers are passed through a furnace to burn the cellulose and to sinter the pTFE. The material is used as a multifilament yarn containing about 30 individual 10-denier fibers. It is somewhat elastic and dark brown due to charred cellulose. In the second method, a sheet of expanded pTFE, such as Goretex, is slit into narrow ribbons, which are twisted (4–10 twists/cm) into a filament. These twisted ribbons can be made to a very high tensile modulus. One problem with these twisted ribbons is that they tend to lay flat, thereby increasing the area that is shielded from the electrical current. Monofilaments with a circular cross-section have been introduced recently to minimize this shielding effect.

In converting these filaments into a fabric, one faces a number of conflicting requirements: given a certain minimum fiber cross-section per cm to achieve the desired strength, these can be arranged as a large number of fine filaments or a lower thread count of thicker threads. From a standpoint of minimizing the obstruction of the electrical current by non-conductive threads, a small number of thicker threads is more desirable. However, there are some limitations imposed by other considerations:

(1) From a cost standpoint and to minimize electrical resistance, it is desirable to use a fairly thin layer of ionomer. Such a layer is more difficult to anchor securely to a few thick threads. Also, the "window" area between threads is unsupported and a thin ionomer layer may not be strong enough for a large "window".

(2) The low coefficient of friction of pTFE makes very open weave patterns difficult to handle: the threads tend to shift even under a modest force. In a plain weave, a minimum thread count of 15/cm is required for reasonable stability. A leno weave pattern is one approach to stabilize an open weave pattern: it uses a double warp. After each fill is inserted between this double warp, the two warp threads are twisted around each other 360° to lock in the fill. With a leno weave pattern, a count of 10/cm is practical.

A thread count as low as 3/cm can be stabilized using sacrificial fibers: the threads of the permanent pTFE reinforcement are alternated with an even number (4 and 8 are commonly used) of "sacrificial" fibers in a plain weave pattern. These fibers could be made of viscose rayon, polyester or some other fiber of poor chemical stability. They will stabilize the fabric temporarily and are destroyed after lamination; to a large extent already during hydrolysis [22]. For instance a fabric with a pTFE thread count of 3/cm using 8× sacrificial fibers would have 24 sacrificial threads per cm for a total thread count of 27/cm. Such a fabric is quite stable, particularly with the higher coefficient of friction of the sacrificial fibers. After chemical destruction, these fibers leave open channels, which can be filled with electrolyte. Because the conductivity of the typical liquid electrolyte is about 10 times higher than that of the ionomer, very conductive membranes can be obtained in this manner. If in a leno weave, the pair of warp filaments consists of a permanent and a sacrificial filament and locks in only every other fill, the weave will convert to a very open plain weave on destruction of the sacrificial filament. In this construction, if the permanent thread of the double warp goes *over* the fill that **is** locked in (the sacrificial thread therefore goes under this fill), the twisted double warp has to go *under* the fill that is **not** locked in (Nafion 2002).

In the lamination step the fabric is fused into a film, or a bifilm for products containing a barrier layer, of precursor polymer(s). This must be done in a way that at least one surface (the barrier layer in the case of a bifilm) of the film is not punctured. To achieve this, particularly for films that are thinner than the fabric at the "knuckles" where warp and fill threads cross, vacuum lamination is very effective (Fig. 3.2): The fabric is placed on a release paper and the precursor film with the barrier layer facing up is placed on top of the fabric. Vacuum is applied from underneath the paper while heat is applied from both above (radiant heat) and below. The melting precursor polymer is forced by the atmospheric pressure down into the window area and underneath the threads. Over most of the area, the polymer flowing under the threads from the two sides actually fuses together, encapsulating the fabric. However, some of the "knuckles" will remain exposed on the underside of the laminate.

The barrier layer would typically have a higher melt viscosity than the base polymer. This helps in maintaining an unbroken "skin" over the fabric, which is essential for this layer to perform its function. In the absence of a barrier layer, a surface hydrolysis can perform a similar function by providing a non-fusible skin of sodium or potassium form polymer [23].

The pores in the release paper must be small enough to prevent the molten polymer from entering. Ordinary paper is much too open for this. Some coating such as those used in offset printing paper is needed to achieve the necessary density.

It can be seen that the two surfaces of the laminate are quite different; the top surface is referred to as the "cathode" side of the membrane, the bottom surface as the "anode" side. This is because the unbroken top skin, with or without barrier layer, is used to stop the electrically driven migration of anions from the catholyte to the anolyte (see Fig. 5.1: NaCl electrolysis). While in most applications the cathode side of the membrane should face the cathode, there are exceptional cases where a "reverse orientation" is preferred (see Section 5.1.6).

It can also be seen that anolyte can enter into the voids within the threads, and particularly into voids created by the use of sacrificial fibers. The anolyte can then seep along these channels and leak out at the edges of the membrane (Fig. 3.3).

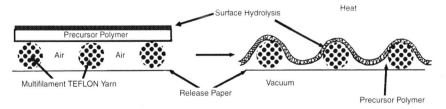

Figure 3.2 Vacuum lamination.

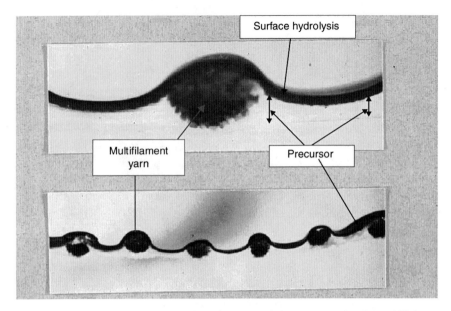

Figure 3.3 Cross-section of a surface-hydrolyzed and vacuum-laminated film after staining.

3.5   Hydrolysis and Acid Exchange

After the precursor polymer has been fabricated to the final physical shape, it is hydrolyzed using a hot solution of sodium hydroxide or potassium hydroxide, optionally also containing some dimethyl sulfoxide (DMSO) to increase the rate of hydrolysis. A typical composition may contain 15% KOH and 30% DMSO. At higher concentrations, phase separation occurs. Some of the DMSO is retained in the final product even after extensive rinsing. If the introduction of a sulfur compound is of concern, for instance as a catalyst poison, then an alkoxy alcohol, such as 1-methoxy isopropanol, can be used [24]. Sodium hydroxide is less expensive than potassium hydroxide and has the advantage that the resulting sodium ion form of the ionomer can be more easily converted to the free acid form. However, DMSO has a more limited solubility in solutions of sodium hydroxide and the rates of hydrolysis that can be achieved in this system are therefore limited. Furthermore, the by-product NaF is not very soluble in the reaction mixture, and can precipitate inside the ionomer film. This is particularly true for fast hydrolysis of thick films.

Film and membrane products are hydrolyzed in a continuous process using a series of heated tanks. The hydrolysis advances with a sharply defined boundary line, as determined by slicing and staining a cross-section. Contact time required at 80°C is 30–60 min for the typical thickness. At 25°C, contact time of 50 h and longer may be required. Even when slicing and staining of a sample hydrolyzed at room temperature indicate hydrolysis through the entire thickness, and the sample has attained some degree of conductivity, infrared analysis will often reveal some unhydrolyzed precursor groups, which will only react at elevated temperatures. The caustic tanks are followed by one or two tanks filled with water to rinse off excess caustic. Elliott *et al.* have studied the hydrolysis reaction by X-ray scattering and *in situ* atomic force microscopy [e-polymers 2001, no. 022].

Those products containing a carboxylate barrier layer, and which represent the largest volume of perfluorinated ionomer products are not subjected to an acid exchange, which would render the carboxylate layer non-conductive. These products are either sold wet, in order to simplify the installation in a cell or treated with a non-volatile swelling agent such as diethylene glycol or sold after drying.

Other products are acid exchanged using nitric or hydrochloric acid. Nitric acid has the advantage that it can be used in stainless steel equipment. To achieve the desired 95% conversion to the acid form in a single-step exchange operation, the utilization of acid is only about 1–2%. That means, an ionomer film, which is 95% in the hydrogen ion form, is in equilibrium with an acid solution in which the ratio of potassium to hydrogen ions is 0.01 or sodium to hydrogen ions is 0.02. A better acid utilization can be achieved in a multistage exchange operation.

After the acid exchange, the film is rinsed with deionized water, usually also in a multistage operation. The final step is drying of the product using warm air.

Unreinforced films of less than 50 μm thickness lack the mechanical strength to pass through the various treatment tanks without an occasional breakage. For those products, film casting (Section 3.6.1) is an attractive alternative.

3.6   Finishing and Testing

Some testing is performed at the precursor stage because at that point most out-of-spec material can be recycled. EW and melt flow are determined at the pelletized precursor stage and most out-of-spec material can be recovered by blending and re-extrusion. Precursor film is visually inspected and tested for thickness. Again, unacceptable materials can be re-extruded.

The hydrolyzed film or laminate is again inspected visually and tested for leaks. An automatic leak tester applies a differential air pressure to the continuous web as it is unrolled and checks for air flow through the sheet. If a leak is detected, the machine stops and an alarm rings. Any defect detected at this stage would have to be eliminated by cutting and would represent a loss of material.

While leak testing by the manufacturer is done continuously on roll goods, a need exists for the end user to check individual sheets for leaks. For instance, in a stack or electrolyzer consisting of several individual cells, a leak may have developed requiring disassembly of the stack and determining which membrane is leaking and where. Quiver Ltd. in Milan, Italy, has developed equipment for this purpose (Fig. 3.4). The membrane is placed on a vacuum table and an arm equipped with sensors traverses the sheet at a rate of 1 m/min (Fig. 3.5, steps 1–3). When one of the sensors detects a leak, the arm stops, an alarm sounds and the sensor that had detected the leak is indicated by a light. The exact location of the leak can then be determined by a manual probe (step 4). Pinholes as small as 0.1 mm can be detected by this method. The test tables are available in four different sizes, all 131 cm wide and with a length of 82, 157, 246 and 300 cm. Quiver also offers equipment to repair leaks by fusing a patch of Nafion 324 over the affected area. The patch must be placed with its anode side against the anode side of the membrane for proper bonding.

The equipment works by thermal impulse, which means that the patch is assembled under the heat sealer at room temperature, pressure is applied by a pneumatic piston and a burst of electric heat raises the temperature of the repair area to the bonding temperature and then allows for rapid cooling (see Chapter 10).

Figure 3.4 Leak test table [25].

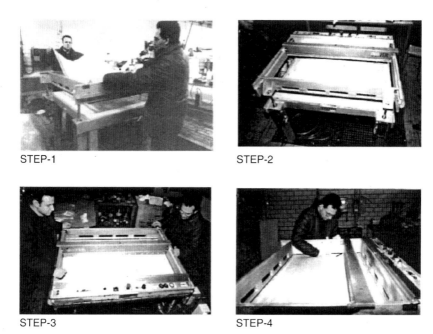

STEP-1   STEP-2

STEP-3   STEP-4

Figure 3.5 Leak testing [25].

3.7 Liquid Compositions

Some fluorinated ionomers dissolve in polar organic solvent or mixtures of such solvents with water under atmospheric conditions. Such polymers include partially fluorinated ionomers based on sulfonated poly trifluoro styrene as well as perfluorinated ionomers of very low EW. Using methanol or ethanol at their atmospheric b.p., the maximum EW for solubility is about 900 for polymers of the Nafion structure and about 750 for the DOW polymer. Perfluorinated ionomers of the more typical EW of 1000–1100 require higher temperatures for dissolution, about 240°C for 1100 EW Nafion [26]. The resulting products are often referred to as "solutions", although they may not necessarily be true solutions of individual polymer molecules but rather agglomerations of several molecules in a micelle containing a hydrophobic core of mostly fluoro polymer backbone surrounded by a hydrophilic shell of ionic groups.

Such a micelle is stabilized by negative charges similar to a soap micelle (see Section 4.3). The addition of a fluoro oxy alkylene oligomer of an MW of 350–500 not only reduces the temperature necessary for dissolution, but also results in smaller micelles [27].

The above procedures use a mixture of water and polar organic solvents, typically lower aliphatic alcohols to convert the ionomer to liquid compositions. More recently, Curtin and Howard Jr. suggested the use of water alone at a temperature of 270°C to obtain true solutions, which means that micelles consisting of a single ionomer molecule. The size of these micelles ranges from 2 to 30 nm [28].

Liquid compositions can be used to impregnate a porous matrix, particularly one made of expanded pTFE, such as Goretex® [23,29,30]. Such composite membranes are offered under the name of Goreselect® by W.L.Gore & Associates. Daramic a battery separator based on porous polyethylene has been impregnated with liquid compositions of Nafion for use in a vanadium redox battery [31].

Another important application for liquid compositions of fluoro ionomers is as a binder in catalyst inks for fuel cells (see Section 5.2).

3.7.1 Film Casting

Particularly for films thinner than about 50 μm, solution casting of a fluorinated ionomer in the final free acid form offers some advantages compared to extrusion of the precursor form. A major advantage is that a backing film can support the product and that the limited strength of a very thin film is therefore of no concern. DuPont, 3M and Ion-Power use this process on a commercial scale. It is described by DuPont as follows (Fig. 3.6):

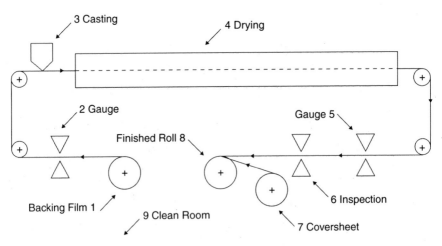

Figure 3.6 Casting an ionomer film from solution [32].

A backing film {1} is measured for thickness {2}. Polymer dispersion is applied {3} to the base film, and both the materials enter a dryer section {4}. The composite membrane/backing film is measured for total thickness {5}, the membrane thickness being the difference from the initial backing film measurement. The membrane is inspected for defects {6}, protected with a cover-sheet {7}, and wound on a master roll {8}. The entire operation is carried out in a clean room environment {9}. This process has several key advantages: (1) pre-qualification of large dispersion batches for quality (e.g., free of contamination) and expected performance (e.g., acid capacity); (2) increased overall production rates for H^+ membrane from solution casting as compared to polymer extrusion followed by chemical treatment and (3) improved thickness control and uniformity, including the production capability of very thin membranes (e.g., 12.7 μm).

3.8 Fluorinated Ionomers with Phosphonic or Sulfonyl Imide Functional Groups

The synthesis of a perfluorinated ionomer containing phosphonic acid groups has been reported in the literature [33]. The synthetic route starts with methyl perfluoro 5-oxa heptenoate (the Flemion® precursor, MW = 306), and uses tetraethyl pyrophosphite to introduce the phosphonic acid group. Burton et al. prepared three different phosphonic acid monomers and polymers, including the Nafion and Dow analog [34]. It was hoped that this functional group would provide better conductivity under low humidity conditions, in analogy to concentrated phosphoric acid, which is fairly conductive due to self-ionization. However, while these polymers had good thermal stability, their conductivity was poorer than that of the corresponding sulfonic acid polymer.

Perfluorinated sulfonyl imides are very strong acids, even stronger than the corresponding sulfonic acids.

Polymers containing sulfonyl imide functional groups have been studied extensively by DesMarteau.

The polymer was obtained by an aqueous copolymerization of a perfluorinated vinyl ether containing a sulfonyl imide sodium salt functional end-group, with TFE [35,36]:

$$CF_3SO_2\overset{\ominus}{N}SO_2CF_2CF_2OCFCF_2OCF=CF_2$$

$$|$$

Na+ CF_3

The resulting polymer is analogous to Nafion, except for the different functional group. Also, because the ionomer is made directly without the use of a precursor, the aqueous polymerization is self-emulsifying: the monomer acts as a surfactant and the micelles are stabilized by their ionic charge. The conductivity of this polymer is similar to that of Nafion.

3.9 Partially Fluorinated Ionomers

Perfluorinated ionomers are fairly expensive compared to other perfluorinated polymers. This is due to three factors:

(1) The more expensive synthesis of the comonomer used.
(2) The much larger amount of this expensive comonomer needed (typically 40–50% by weight).
(3) The relatively small scale of manufacture.

As a result, there have been numerous attempts to replace these expensive polymers with less expensive partially fluorinated ionomers. In the first approach, non-fluorinated monomers, such as styrene, are grafted on a commercially available pTFE or FEP film. This approach takes advantage of the economics of scale of film fabrication. In the second approach, a less expensive partially fluorinated monomer is used.

In general, the chemical stability of the partially fluorinated ionomers is inferior to that of fully fluorinated ionomers.

3.9.1 Grafting of Non-fluorinated Monomers on Fluorinated Films

Radiation grafting of styrene on FEP or pTFE films followed by sulfonation has been used to obtain partially fluorinated ionomer films. Styrene also has been grafted on "NAFION®" in supercritical carbon dioxide using an azo initiator [37]. Radiation grafting of styrene on cross-linked pTFE is described in Ref. [38]. Radiation grafting of styrene on poly vinyliden fluoride (PVDF) and PVDF/HFP copolymers has been reported in Ref. [39]. In all these cases, the graft polymer is subsequently sulfonated. The stability of films obtained by radiation grafting of styrene on FEP can be improved by the addition of divinyl benzene (DVB) as a cross-linking agent to the grafting solution [40]. As an example, a 25-μm-thick FEP film (FEP 100A, DuPont) was irradiated by an electron beam with a dose of 3–5 kGy, followed by introduction into a stainless steel reactor containing the grafting solution. The grafting solution consisted of styrene containing 0–20% DVB dissolved in a isopropanol/water mixture. The films were then washed with toluene and dried. The degree of grafting as determined by the weight increase was 17–21%. The film was sulfonated using a 2 vol.% solution of chorosulfonic acid in methylene chloride. Complete sulfonation was verified by titration. Without any DVB, the EW for the degree of grafting given can be calculated as 800–680. For DVB containing grafts, these numbers would be slightly higher. The thickness of the DVB-containing films was 33–35 μm, and without DVB it was 39 μm.

The chemical stability of such graft polymers is of course limited, particularly by the oxidative instability of the C–H bond α to the aromatic ring (compare the easy conversion of cumene to phenol and acetone through a peroxide intermediate). Some efforts have been made to stabilize these graft polymers through the elimination of the α-C–H bond. α, β, β-trifluoro styrene is a logical candidate (see Section 3.8.2). α-methyl styrene alone results in poor grafting yield, however in combination with acrylo nitril, reasonable results have been obtained [41] (see also Section 3.8.3 [42]).

Stone *et al.* described the radiation grafting of trifluorovinyl aromatics on a base film made of an ethylene/TFE copolymer (TEFZEL®) [43].

3.9.2 Polymerization of Partially Fluorinated Monomers

The most commonly used non-fluorinated ion exchange resins are based on cross-linked polystyrene. In this structure, the benzylic carbon (the backbone carbon to which the aromatic ring is attached) is most sensitive to oxidation. Even minor cleavage of the backbone results in drastic loss of MW and/or cross-linking. In contrast, an attack on the aromatic branch would result in only a small loss of ion exchange capacity. α, β, β-trifluoro styrene therefore has been used as a monomer to yield a polymer with a perfluorinated backbone and improved oxidative stability. Sulfonic acid groups are introduced by sulfonation of the polymer. Sulfonation is usually complete and such polymers exhibit excessive swelling in water and brittleness when dry. Control of the degree of sulfonation and thereby swelling behavior was made possible by the introduction of a comonomer with reduced reactivity in sulfonation.

Ballard Power Systems has commercialized these ionomers. A comonomer, such as *p*-fluoro- or *m*-trifluoro methyl-α, β, β-trifluoro styrene, with reduced reactivity toward sulfonation is used to allow control of the EW. The precursor polymers are soluble in chlorinated solvents, such as chloroform, and are sulfonated in solution using chlorosulfonic acid or a sulfur trioxide/triethyl phosphate complex dissolved in chloroform [44,45].

Table 3.3 Partially Fluorinated Ionomer Films Based on Radiation Grafted pTFE or FEP

Design	Type	Resistance	Selectivity	Capacity	% Water	Thickness
R-1010	Cat./pTFE	0.2–0.5	86 %	1.2 mEq/g	20	50
R-1035	An./pTFE	0.7–3.0	81 %	1.0	10	50
R-4010	Cat./FEP	0.2–1.0	82 %	1.5	25	100–130
R-4035	An./FEP	1–2.5	80 %	1.2	33	80–100

Raipore® membranes prepared by RAI Research Corporation.

An improved synthesis of α, β, β-trifluoro styrene and substituted α, β, β-trifluoro styrenes from 1,1,1,2-tetrafluoro ethane is described in Ref. [46].

3.9.3 Fluorinated Anion Exchange Polymers

Perfluorinated amines do not exhibit any basic properties; perfluoro tributyl amine for instance is insoluble in concentrated sulfuric acid. As a result, fluorinated anion exchange polymers are of necessity only partially fluorinated. They can be prepared by the reaction of di- or polyamines containing at least one secondary and one or more tertiary amino groups (see Section 8.3).

Radiation grafting on FEP has been used to prepare fluorinated anion exchange membranes [42]. Table 3.3 shows anion and cation exchange films based on radiation grafting styrene on films of TFE or FEP. The functional groups are trimethyl benzyl ammonium and sulfonic acid, respectively.

3.10 Composite Materials of Ionomers and Inorganic Oxides

A wide variety of inorganic oxides have been incorporated into fluorinated ionomers in an effort to increase the maximum use temperature, particularly under low humidity conditions. This is important for fuel cell applications, where a higher operating temperature increases the carbon monoxide tolerance and allows for easier dissipation of the waste heat. In direct methanol fuel cells, such composites are reported to reduce the methanol crossover. Thampan *et al.* in a recent publication offered a systematic analysis of the aspects of composite ionomers [47]. Ionomer composites also are useful for catalytic applications.

There are three approaches to incorporate a Dopant or filler into an ionomer: a finely divided solid filler can be preformed, and then added to a liquid composition of the ionomer followed by film casting or other method of fabrication. Or the filler can be formed inside the solid ionomer from an alcoholic solution of a precursor, such as tetraethyl silicate, zirconate or titanate. In the third approach, both the inorganic and the ionomer component are in liquid form, which are mixed and coprecipitated. For example, an ionomer-modified silica gel can be prepared by mixing a solution of sodium silicate with a liquid composition of Nafion, possibly with additional sulfuric acid. With the proper acid-to-base ratio, the initially liquid composition will turn into a gel, which is dried and subsequently leached with dilute nitric or hydrochloric acid to remove any sodium sulfate and convert the ionomer back into the hydrogen form. The advantage of the first approach is that high temperatures can be used in the preparation of the filler and it can be characterized more easily by various analytical techniques. The second approach on the other hand is more likely to yield a composite in which the filler is intimately associated with the ionic sites of the ionomer for a synergistic effect, while to some extent maintaining the cluster morphology of the ionomer. The third approach is likely to create an entirely new morphology.

In the first approach, the filler is simply suspended in a solution of the ionomer in a suitable solvent and the resulting mixture cast on a glass plate utilizing a doctor blade and dried for 15 min at 100°C. The film is released from the glass plate by wetting with DI water and dried. The dried film can then be annealed for 15 min at 170°C in a hydraulic press between sheets of TEFLON® [48]. A particularly attractive filler for the first approach is sulfated zirconium dioxide. This material, which is one of the strongest known solid super acid [49], is obtained by treating zirconium dioxide with dilute sulfuric acid, followed by drying at 100°C to 120°C. The powder is then calcined for 2 h at 600°C in air. The material is then crushed in a mortar or pulverized in a Jet Mill (Laboratory Jet Mill, Clifton NJ) [50].

Another example is the use of titanium dioxide reported recently by Chalkova et al. [51]. The rutile form was obtained as a powder (Kronos 4020) from Kronos, Cranbury, NJ. The material consisted of 3–5 μm aggregates of 0.1–1.0 μm rutile grains and had a surface area of 2.9 m²/g. It was mixed with a 5% solution of Nafion (Aldrich) and sonicated in an ultrasonic bath. A film was cast from this suspension and dried at 80°C. The film was then repressed at low pressure at 150°C for 10 min. Film thickness was about 80 μm. The properties and fuel cell performance of films containing 10% and 20% titanium dioxide were compared with those made without titanium dioxide.

Recently, Zaidi *et al.* reported on composites prepared by casting a mixture containing 10–50% of boron phosphate in a solution of a per-fluoro sulfonic acid ionomer in isopropanol. An increased conductivity, compared to films cast from the unmodified solution, was observed for composites containing 20% or more of boron phospate [52].

In the second approach, a Nafion film was swollen in 90% ethanol for 1 h. The swollen film was then immersed for 10 min in a 5% solution of zirconium *t*-butoxide in ethanol, rinsed briefly in ethanol and dried for 24 h in a vacuum oven at 110°C [53]. Optionally, a zirconium dioxide/Nafion composite, made in a similar fashion from Nafion 112 and a solution of zirconium iso-propoxide in iso-propanol, can be sulfated by heating in 1 M sulfuric acid. At 80% RH, these composites exhibit higher conductivity than the host film [54]. Films prepared by using the second approach are completely transparent, indicating the absence of large filler particles. Other oxides that can be incorporated in ionomer films by this approach include the oxides of silicon and titanium.

As an example for the third approach, Jiang *et al.* added tetraethyl ortho-silicate to a liquid composition of Nafion followed by solution casting. The resulting films were characterized by TGA, XRD, FTIR, SEM-EDX and water uptake, and were investigated in direct methanol fuel cells. Low silica loadings (3% and 5%) helped to inhibit methanol crossover. Proton conductivity decreased with increasing silica content, in spite of the increasing water uptake. The performance of a composite membrane with 5% silica loading was higher than that of a pure Nafion membrane [55]. Ramani *et al.* reported on the effect of the particle size of the filler on the performance of a Nafion/phosphotungstic acid composite. Particle sizes of 1–2 μm were compared with those of 0.03 μm [56].

The properties of these composites as well as their performance in fuel cells will be discussed in Chapter 6 "Fuel Cells and Batteries".

3.11 Remanufactured Membranes

In view of the relatively high price of ionomer membranes, particularly the perfluorinated types, efforts have been made to manufacture such membranes from recovered material. The recovered material could be either manufacturing scrap (remnants, trimmings, off-spec material, etc.) or end-of-life products. The end-of-life products could be used chlor-alkali membranes, which are now available in substantial quantities, or used fuel cell components, particularly MEAs, which may become available in large quantities in the future.

The recovery process yields the valuable components separately: sulfonic polymer, carboxylic polymer and, in the case of fuel cell components, platinum catalyst. Any fabric reinforcement, while being recovered, cannot be reused [57]. It is also hoped that by recovering the individual components separately, a "post-mortem" may help to identify the cause of failure.

References

1. Sherratt, S., in: *Kirk-Othmer Encyclopedia of Chemical Technology*, 2nd edition (A. Standen, ed.), 9: 805–831, Interscience Publishers, Division of John Wiley & Sons, New York, 1966.
2. Hals, L.T., Reid, T.S., Smith, G.H., J. Am. Chem. Soc., **73**, 4054, 1951.
3. West, N.E., US Patent 3,873,630 assigned to DuPont, Mar. 25, 1975.
4. Hauptschein, M., Fainberg, A.H., US Patent 3,009,966 assigned to Pennwalt Chemicals, Nov. 21, 1961.
5. Eleuterio, H.S., Meschke, R.W., US Patent 3,358,003 assigned to DuPont, Dec. 12, 1967.
6. Carlson, D.P., US Patent 3,536,733 assigned to DuPont, Oct. 27, 1970.
7. Putnam, R.E., Nicoll, W.D., US Patent 3,301,893, assigned to DuPont, Jan. 31, 1967.
8. Millauer, H., Schwertfeger, W., Siegemund, G., Angewandte Chemie, Intern. Edit. (Engl.), **24**, 161–179, 1985.
9. Yamabe, M., Arai, K., Samejima, S., Noshiro, M., US 4,116,977 assigned to Asahi Glass, Sep. 26, 1978.
10. Yamabe, M., Munekata, S., Sugaya, Y., Jitsgiri, Y., US Patent 4,127,731 assigned to Asahi Glass, Nov. 28, 1978.
11. Yamabe, M., Munekata, S., Samejima, S., US Patent 4,153,804 assigned to Asahi Glass, May 8, 1979.
12. Ghielmi, A. *et al.*, Paper presented at the Grove Fuel Cell Symposium, Munich, Oct. 6, 2004; Ghielmi, A., Vaccarono, C., Troglia, C., Arcella, V., J. Power Source., **145**(2), 108–115, 2005.
13. Ukihashi, H., Yamabe, M., Miyake, H., Prog. Polym. Sci., **12**, 229–270, 1986. See also Ukihashi, H., CHEMTECH, pp. 118–129, Feb. 1980.
14. Guerra, M.A., US Patent 6,624,328 assigned to 3M Innovative Properties Co., Sep. 23, 2003.
15. Ezzell, B., Carl, W., Mod, W., US Patent 4,358,412 assigned to Dow Chemical, Nov. 9, 1982.
16. Ezzell, B., Carl, W., Mod, W., US Patent 4,337,211 assigned to Dow Chemical, Jun. 29, 1982.
17. Kimoto, K., Miyauchi, H. Ohmura, J. Ebisawa, M., Hane, T., US Patent 4,597,913 assigned to Asahi Kasei, Jul. 1, 1986.
18. England, D., US Patents 4,131,740, Dec. 26, 1978 and 4,138,426, Feb. 6, 1979 assigned to DuPont.
19. Quarderer, G., Katti, S., Polymer Eng. Sci., **33**(9), 564–572, 1993.
20. Apostolo, M., Arcella, V., US Patent 6,555,639 assigned to Ausimont, Apr. 29, 2003.

21. Grot, W.G., US Patent 5,281,680 assigned to DuPont, Jan. 25, 1994.
22. Grot, W.G., US Patent 4,021,327 assigned to DuPont, May 3, 1977.
23. Grot, W.G., US Patent 3,770,567 assigned to DuPont, Nov. 6, 1973.
24. Banerjee, S., Grot, W.G., US Patent 5,310,765 assigned to DuPont, May 10, 1994.
25. Quiver, Ltd., Milan, Italy, Product Literature.
26. Grot, W.G., US Patents 4,433,082, Feb. 21, 1984 and 4,453,991 Jun. 12, 1984 assigned to DuPont.
27. Maccone, P., Zompatori, A., US Patent 6,197,903 assigned to Ausimont, Mar. 6, 2001.
28. Curtin, D., Howard Jr., E., US Patent 6,150,426 assigned to DuPont, Nov. 21, 2000.
29. Bahar, B. *et al.*, US Patent 5,547,551 assigned to W.L.GORE, Aug. 20, 1996.
30. Bahar, B., Hobson, A., Kolde, J., US Patent 5,599,614 assigned to W.L.GORE, Feb. 4, 1997.
31. Tian, B., Yan, C.W., Wang, F.H., J. Membr. Sci., **234**(1–2), 51–54, May 1, 2004.
32. Curtin, D. *et al.*, J. Power Source., **131**, 41–48, 2004.
33. Kato, M., Akiyama, K., Yamabe, M., Reps. Res. Lab., Asahi Glass Co. Ltd., **33**(2), 135, 1983.
34. Kotov, S., Pedersen, S., Qiu, Z., Burton, D., J. Fluorine Chem., **82**, 13–19, 1997.
35. Thomas, B., Shafer, G., Ma, J., Tu, M., DesMarteau, D., J. Fluorine Chem., **125**(8), 1231–1240, 2004.
36. Thomas, B., DesMarteau, D., J. Fluorine Chem., **126**(7), 1057–1064, 2005.
37. Sauk, J., Byon, J., Kim, H., J. Power Source., **132**, 59–63, 2004.
38. Tetsuya, Y. *et al.*, Polymer, **45**(19), 6569–6573, Sep. 3, 2004.
39. Soresi, B. *et al.*, *Solid State Ionic.*, **166**(3–4), 383–389, Jan. 30, 2004.
40. Schmidt, T., Simbeck, K., Scherer, G.G., J. Electrochem. Soc., **152**(1), A93–A97, 2005.
41. Becker, W., Bothe, M., Schmidt-Naake, G., Angew. Makromol. Chem., **273**, 57–62, 1999.
42. Slade, R., Varcoe, J.R., Solid State Ionic., **176**(5–6), 585–597, Feb., 2005.
43. Stone, C., Steck, A.E., US Patent 6,359,019 assigned to Ballard, Mar. 19, 2002.
44. Steck, A.E., Stone, C., Wei, J., US Patent 5,422,411 assigned to Ballard Power Systems, Jun. 6, 1995 and continuations in part 5,498,639, Mar. 12, 1996 and 5,773,480, Jun. 30, 1998.
45. Holdcroft, S., Ding, J., Chuy, C., Stone, C., Morrison, A., US Patent 6,765,027 assigned to Ballard Power Systems, Jul. 20, 2004.
46. Stone, C. *et al.*, US Patent 6,653,515 assigned to Ballard Power Systems, Nov. 15, 2003.
47. Thampan, T., Jalani, N., Choi, P., Datta, R., J. Electrochem. Soc., **152**(2), A316–A325, 2005.
48. Moore, R.B., Martin, C.R., Anal. Chem., **58**, 2570, 1986.
49. Misono, M., Okuhara, T., Chemtech, **23**, 1993. Also J. Mol. Catal., **74**(1–3), 247–256, 1992.
50. Arata, K., Catal. A, **146**, 3, 1996.
51. Chalkova, E., Pague, M., Fedkin, M., Wesolowski, D., Lvov, S., J. El. Chem. Soc., **152**(6), A1035–A1040, 2005.

52. Zaidi, J., Rahman, S.U., J. Electrochem. Soc., **152**(8), A1590–A1594, 2005.
53. Liu, P., Bandara, J., Lin, Y., Elgin, D., Allard, L., Sun, Y., Langmuir, **18**, 10398, 2002.
54. Choi, P., Jalini, N., Datta, R., J. Electrochem. Soc., **152**(8), A1548–A1554, 2005.
55. Jiang, R., Kunz, R., Fenton, J., J. Membrane Sci., **279** (1–2), 506–512, Aug., 2006
56. Ramani, V., Kunz, R., Fenton, J., J. Membr. Sci., **266**(1–2), 110–114, 2005.
57. Grot, S., Grot, W., US patent application 20050211630, assigned to Ion-Power, Sept. 29, 2005.

4 Properties

Much of the early literature describes the properties of Nafion®. More recently, reliable information about the new Hyflon-Ion® polymer has become available [5,9]. This polymer is based on the same monomer as the experimental Dow polymer, which has been discontinued.

The comparison of ionomers of different length of the side chain is somewhat difficult. In the literature, this comparison is usually made on the basis of equal equivalent weight (EW). This is neither entirely reasonable nor practical, as can be seen by the Nafion–Dow comparison: It assumes that removing 166 weight units from the branches (166 is the weight of the extra HFPO unit in Nafion) and replacing it with 1.66 extra polytetrafluoroethylene (pTFE) units in the backbone should have no effect on the properties. This of course is not true. If the comparison is done on the basis of identical comonomer ratios, then the EW scale is displaced by 166 units (1046 EW Nafion would be compared to 880 EW Dow polymer). This comparison assumes that the extra hexafluoro propylene epoxide (HFPO) unit in the branch should not affect the properties of the polymer. This is also not the case; the extra HFPO unit decreases crystallinity, enhances the formation of ionic clusters and in general "plasticizes" the polymer similar to the effect of additional comonomer. From an empirical standpoint, a displacement of about 240 EW units (1046 EW Nafion compared to 806 EW Dow) would put the two polymers on a comparable basis. On this basis, the Hyflon-Ion polymer may exhibit a superior balance of properties if other polymer qualities, such as EW distribution, molecular weight (MW) and MW distribution, and end group stability, can be maintained. It appears that the Dow polymer, dealt earlier, may have been deficient in this respect.

4.1 Properties of the Precursor Polymers

It should be emphasized that the precursor forms are not ionomers and have properties completely different from the ionic forms. The precursor polymers are hydrophobic, non-conducting, inert materials that resemble a highly plasticized (because of the high comonomer content) TFE copolymer. They are readily melt-fabricable. When marked with a felt point pen, the ink beads up (easy distinction from the ionic forms).

Table 4.1 Uniaxial Draw of Precursor at 70°C

Draw ratio		
MD	**TD**	**Planar strain**
1.68	0.69	1.16
2.00	0.64	1.28
3.08	0.54	1.66
5.36	0.38	2.4
5.68	0.38	2.16
7.72	0.33	2.55

Source: Ref. [1].

Extruded films of the Nafion precursor polymer are soft and pliable and tend to stick to themselves, particularly at low EW. They are easily stretchable at room temperature or slightly elevated temperature with some shrinkage in the transverse direction (TD) (Table 4.1). Elongation at break is about 340% at room temperature and about 1000% at 70°C for 1100 EW Nafion precursor film (Figs. 4.1–4.3) [1].

Table 4.1 shows samples of Nafion 1100 EW precursor form were compression molded at 265°C. They were stretched at 5 cm/min crosshead speed and the length (machine direction, MD) and width (TD) between markings on the sample measured 1 week after release from the tensile tester. The draw ratio is the length and width after the test divided by the original dimensions. The planar strain is the product of MD and TD draw ratios.

The crystallinity of the Nafion precursor polymer has been reported earlier as about 15–20% [2–4]. A more recent publication [1] suggests values of 3.8–10.5%. The crystallinity decreases with decreasing EW and disappears below 965 EW for the Nafion precursor. The crystallinity is responsible for a broad endotherm from about 130–250°C. Melt-fabrication of the precursor polymer is typically done at about 280°C. The glass transition temperature is below 10°C.

For the Hyflon-Ion precursor polymer, crystallinity disappears below 700 EW. The differential scanning calorimetry (DSC) curve for 850 EW Hyflon-Ion is shown in Fig. 4.4. The glass transition temperature for this precursor polymer ranges from 5°C for an EW of 510 to 50°C for an EW of 1180.

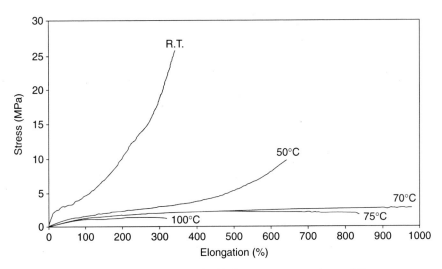

Figure 4.1 Stress and strain curves for Nafion precursor polymer [1].

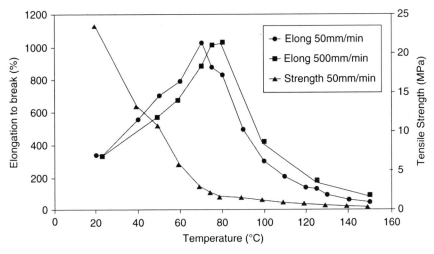

Figure 4.2 Elongation at break and tensile strength of Nafion precursor film [1].

The heat of second fusion of samples of Hyflon-Ion and Nafion precursors are shown in Fig. 4.5. (Second fusion means that the samples had been heated previously once for 15 min to 350°C.)

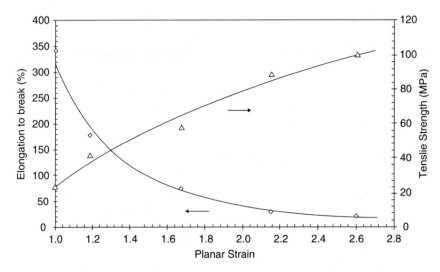

Figure 4.3 Tensile properties of pre-stretched Nafion precursor polymer [1].

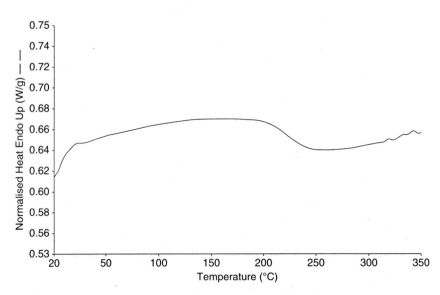

Figure 4.4 DSC curve of Hyflon precursor polymer 850 EW [5].

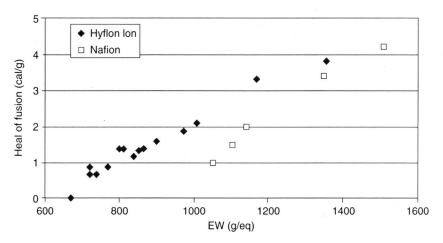

Figure 4.5 Heat of second fusion of Hyflon-Ion and Nafion [5].

4.2 Properties of the Ionic Forms

The ionic forms, with the exception of those containing large organic cations, are noticeably stiffer than their precursor polymers because of ionic cross-linking. The glass transition temperature of Nafion 1100 EW in the sulfonic acid form is about 110°C, and 165°C for the Dow polymer.

The dominant property of fluorinated ionomers is their ability to absorb large quantities of water or polar organic solvents. Other properties are then very much a function of the polymer's water or solvent content. With increasing water or solvent content the dimensions, conductivity and permeability of the ionomer increases, and the strength and the ability to reject anions decreases. The water content in turn is controlled by the following factors: (1) the ambient humidity (or availability of water) conditions; (2) the EW; (3) the nature of the cation and (4) the pretreatment history. These will be discussed below.

The "standard" pretreatment used to establish a reproducible history for test samples is boiling for 30 min in water. This pretreatment, however, will not completely "erase" the effect of previous drying at temperatures above 120°C.

Mechanical properties of 1100 EW Nafion are shown in Table 4.2 for products based on extruded precursor film and solution cast film [6]. The tear resistance is thickness dependant; the values given are for Nafion 112 (50 μm thick).

Table 4.2 Properties of Nafion® Perfluorosulfonic Acid Membranes

Nafion type	Typical thickness (µm)	Basis weight (g/m²)
N 211 (solution cast)	25.4	50
N 112	51	100
N 212 (solution cast)	50.8	100
N 1135	89	190
N 115	127	250
N 117	183	360
N 1110	254	500

Property	N 112, 1135, 115, 117, 1110	N 211	N 212
Tensile modulus, 50% RH (MPa)	249	284	258
In 23°C water	114		
In 100°C water	64		
Maximum tensile strength, 50% RH (MPa)	43 MD, 32 TD	23 MD, 28 TD	32 MD and TD
In 23°C water	34 MD, 26 TD		
In 100°C water	25 MD, 24 TD		
Elogation at break, (%) 50(%) RH	225 MD, 310 TD	252 MD, 311 TD	343 MD, 352 TD
In 23°C water	200 MD, 275 TD		
In 100°C water	180 MD, 240 TD		
Tear Resistance, initial (g/mm) 50% RH	6000 MD and TD		
In 23°C water	3500 MD and TD		
In 100°C water	3000 MD and TD		
Tear Resistance, propagation (g/mm) 50% RH	>100 MD, >150 TD		
In 23°C water	92 MD, 104 TD		
In 100°C water	74 MD, 85 TD		
Water Uptake at 100°C (%)	38	50	50
Linear Expansion in 100°C water, (%)	15	15	15

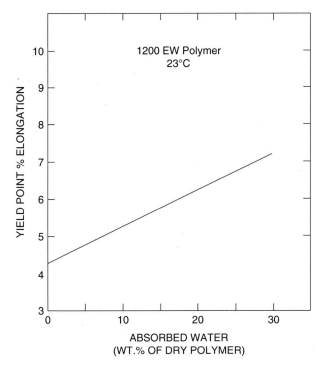

Figure 4.6 Elongation at yield [8].

Ionomers in sheet form are generally produced in a continuous process, in which the material is subject to a stress in the MD. As a result, the material is pre-stressed in the MD and the mechanical properties and expansion values are slightly different in the MD and the TD.

The mechanical properties of solution cast films (N 211 and N 212) at 50% RH are comparable to those of products based on extruded precursor film. However, water uptake of solution cast film at 100°C is higher and one may expect that the mechanical properties under these conditions are somewhat poorer.

Elongation at yield, tensile modulus and tensile strength as a function of water content for Nafion 1200 EW in the hydrogen form are given in Figs 4.6–4.8.

Table 4.3 compares the tensile properties of Hyflon-Ion of 800 and 900 EW with those of Nafion 115 (1100 EW) and Gore-Select; Table 4.4 gives these data for samples that had been pretreated for 30 min in 100°C water.

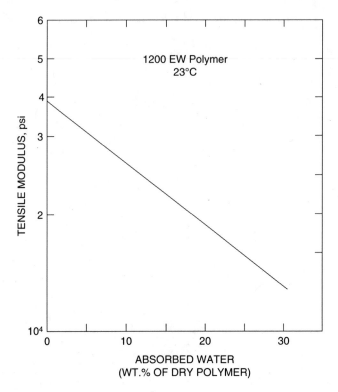

Figure 4.7 Tensile modulus [8].

Other properties of some perfluorinated ionomer membranes are listed in Table 4.5. They are based on the monomers listed in Table 3.1. These membranes are older types for which data are more readily available. The trend in chlor-alkali membranes has been to thinner and lower EW barrier layers to reduce electrical resistance. This trend can be seen in comparing the Aciplex membrane with the older Nafion 90209. The general descriptions of these membranes are:

Nafion 117: A 180-μm thick film of 1100 EW sulfonic acid polymer without reinforcement. The main application for this material is in fuel cells.

Nafion 324: A main layer of 125 μm, 1100 EW sulfonic acid polymer reinforced with a 10 × 10/cm leno weave of 250 denier brown multifilament yarn. On the cathode side of the membrane is a 25 μm barrier layer of 1500 EW sulfonic acid polymer for improved anion rejection. Recent production of this material appears to be calendered to reduce the overall thickness, but this information is not confirmed by the manufacturer.

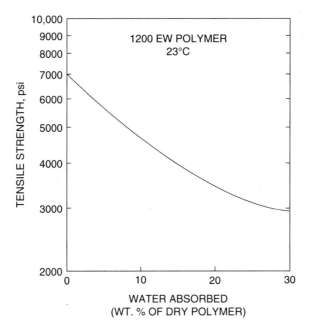

Figure 4.8 Tensile strength [8].

Table 4.3 Stress and Strain at Break for Hyflon-Ion–FM Membranes and Other Commercial Membranes, Measured at RH = 50% 23°C

Membrane	Stress (Mpa)		Strain (%)	
	MD	TD	MD	TD
Hyflon-Ion–FM EW = 800	23	20	130	120
Hyflon-Ion–FM EW = 900	30	23	90	140
Nafion 115	30	26	119	188
Gore-Select[36]	34	24	–	–

Membranes conditioned under the measurement conditions: MD, machine direction; TD, transverse direction.
Source: Ref. [9].

N 324 is a general purpose industrial membrane particularly for applications requiring better anion rejection than N 424.

Nafion 417: A single layer of 180 μm 1100 EW sulfonic acid polymer reinforced with 16 × 16/cm plain weave of 400 denier brown multifilament yarn (essentially reinforced N 117). N 417 has been replaced by Nafion 424.

Table 4.4 **Stress and Strain at Break for Hyflon-Ion–FM Membranes and Other Commercial Membranes, Measured at RH = 50% 23°C After a 30-min. Boil in Water**

Membrane	Stress (Mpa)		Strain (%)	
	MD	**TD**	**MD**	**TD**
Hyflon Ion–FM EW = 800	14	15	100	125
Hyflon Ion–FM EW = 900	22	17	70	110
Nafion 115	23	18	85	110
Gore-Select[36]	32	17	–	–

Membranes prepared by soaking in water at 100°C for 30 min. MD, machine direction; TD, transverse direction.
Source: Ref. [9].

Table 4.5 **Properties of Some Fluorinated Ionomer Films and Membranes**

	N 117	**N 324**	**N 417**	**N 954**	**N 90209**	**F 4221**
Ionic form	H	H	H	K	K	Na
Thickness (μm)	180	300	400	290	300	
Weight (g/m^2)	340	480	500	422	420	
Density* (g/cm^3)	1.95					
Resistance in 24% NaCl, (Ωcm^2)	1.4	4.5	4.1	2.6	2.6	
Linear expression, MD/TD (%)						
Water	16	5/6	9/12			
2% NaOH	12	4/5	7/11	8/8	8/8	7/7
25% NaCl	8	2/3	5/8			
Thermal conductivity (J/cm °C h)	6.5					
Tensile strength (g/cm)						
50% RH (maximum)	4400	7900	10500			6260
Wet (maximum)	2600	7900	10500	5120	5120	5130

*The density given here is at 50% RH; the density of dry Nafion is 2.10 g/cm^3.

Nafion 424: uses the same leno weave as N 324. N 417 and N 424 are general purpose membranes for applications where durability and strength are more important than anion rejection.

Nafion 350 and 450 are now discontinued versions of Nafion 324 and 424, respectively. They used a white leno weave of a twisted ribbon of expanded pTFE as a reinforcement.

Nafion 954: A main layer of 125 μm 1080 EW sulfonic acid polymer reinforced with a 6.5 × 6.5/cm plain weave of 200 denier twisted ribbon (4 twists/cm) of expanded pTFE, interwoven with 26 × 26/cm of a sacrificial fiber. A 38 μm barrier layer of 1050 EW carboxylic polymer plus a gas release coating on the cathode side.

Nafion 90209: The same as N 954 except without the gas release coating.

Aciplex F 4221 (Asahi Chemicals): A main layer of 125 μm of 1030 EW sulfonic acid polymer reinforced with an 8 × 8/cm plain weave of 200 denier twisted ribbon (8 twists/cm) of expanded pTFE (no sacrificials). A 20 μm barrier layer of 1010 EW carboxylic polymer. A gas release coating is applied to both sides. It appears that this coating has been applied to the precursor film before lamination.

4.2.1 Swelling in Water and Other Solvents

Fluorinated ionomers swell, and in some cases dissolve, in polar solvents. The properties and morphology of these polymers are dominated by this swelling process. In some cases a mixture of solvents is a more effective swelling agent than either component alone (Fig. 4.9). Extensive data on swelling of Nafion 117 in organic solvents and solvent/water mixtures has been published by Gebel *et al.* [7], see also Table 4.6 taken from this reference. A high degree of swelling, particularly at elevated temperature, may result in partial dissolution of the ionomer. Siroma reported that at 80°C in methanol or methanol/water mixtures containing more than 80% methanol, more than 30% of a Nafion film is dissolved [11].

As a general rule, an organic liquid is likely to swell a perfluorinated ionomer, if this liquid is completely or at least partially miscible with water. In the case of partial miscibility, a combination of this liquid with water is likely to be a more effective swelling agent than either component alone.

An extremely effective swelling agent for Nafion is a mixture of approximately equal amount of water and trimethyl phosphate. Unfortunately, this mixture is subject to slow hydrolysis. Related to this is the swelling in dimethyl methylphosphonate. This solvent is reported to swell only the fluoroether part of Nafion [14]. In general, some derivatives of phosphoric acid, such as hexamethyl phosphotriamide, and also phosphoric acid itself exhibit a unique swelling power for Nafion.

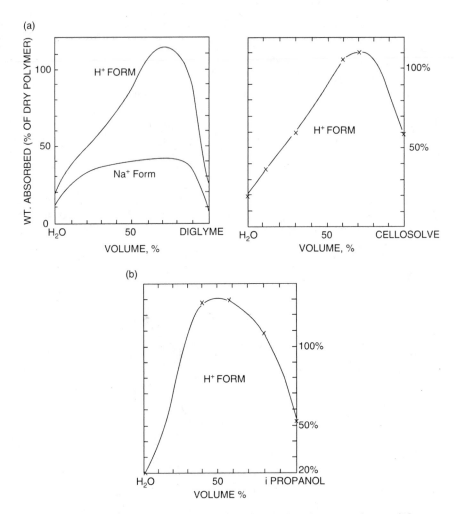

Figure 4.9 Swelling of Nafion in a mixture of organic solvents and water [8].

The swelling of an ionomer in a solvent or solvent mixture is a function of several factors:

1) The nature of the solvent or solvent mixture as discussed above.
2) The composition and particularly the EW of the ionomer.
3) The nature of the counter-ion.

Table 4.6 Swelling of NAFION in Various Solvents

Solvent	$\Delta l_1/l1$ (%)	$\Delta l_1/l$ (%)	$\Delta e/e$	$\Delta V/V$ (%)	N_{SO3}
Water	14	10	14	43	13.1
Methanol	51	36	50.5	209	28.4
Ethonal	45	32	47	181	17.1
2-Proponal	32	22	33	114	8.2
1-Butanol	45	30.5	45	174	10.5
1-Pentanol	44.5	31.5	45	176	8.9
Glycerol	17	12	16.5	52	4.0
Ethylene glycol	28	20	28	97	9.5
Formamide	28.5	20.5	28.5	99	13.7
N-methylformamide	51	37	54	219	20.6
N,N-dimethylformamide	53	38	52	221	15.8
N,N-diethylformamide	53	37	53.5	222	10.9
N,N-dimethylacetamide	68	49	70	326	19.3
N,N-diethylforamamide	74	53	75	366	16.0
Trimethylphospate	40	28	49	167	7.8
Triethylphospate	46	33	66	222	7.2
Tributylphosphate	65	46	91	360	7.3
Hexamethylphosphotriamide	98	73	143	732	26.2
Dimethylsulphoxide	45.5	32.5	45	180	13.9
N-methylpyrrolidinone	53	37	54	223	12.7
Dimethylsulphoxide	45.5	32.5	45	180	13.9
N-methylpyrrolidinone	53	37	54	223	12.7
Cyclohexanone	21	15	20.5	68	3.6
2-Ethoxyethanol	36	25	37	133	7.5
Tetrahydrofuran	20.5	14.5	20	66	4.7
Propylene carbonate	18	13	18.5	58	3.7
Butyl acetate	15.5	11	17	50	2.1
Dioxane	15	11	16	47	3.0
Pyridine	36.5	26	38	137	9.4
Hydrazine	14	10	15	44	7.6
Acetonitrile	16.5	11	16	50	5.3

Source: Ref. [7].

4) In exposure to other than liquid solvents, the vapor pressure (or humidity in case of water).
5) The temperature of exposure.
6) The hydration history.

There are different ways to express the solvent uptake of an ionomer. In the case of water, it can be expressed as percentage based on dry ionomer, or grams or moles of water per gram of dry ionomer. From the last expression, mol/g, other expressions are derived:

Moles of water per ionic group = EW × mol/g

"Hydration product" = $(EW)^2$ × mol/g

The hydration product is an expression developed to differentiate the hydration properties of polymers of different structures (US Patent 4,358545).

The EW of the ionomer is of course an important factor controlling solvent uptake. As the EW decreases, the concentration of ionic groups increases and therefore the interaction with the solvent. The structure of the polymer, however, is also a factor. For instance, Nafion will swell more than a Dow polymer of equal EW, as illustrated in Table 4.7 (values in moles of water per exchange site).

Instead of comparing the two polymers at the same EW, they could also be compared at the same comonomer ratio. Because the MW of the Dow monomer is 166 units lower than that of the Nafion monomer (see Table 3.1), the EW of the Dow polymer is also 166 units lower than that of Nafion of equal comonomer ratio. Therefore, 1000 EW Nafion has

Table 4.7 Moles of Water Absorbed per Exchange Site for Nafion and Dow Polymers

Nafion		Dow	
EW	mol	EW	mol
850	35.1	800	36.8
1000	25.4	830	21.8
1200	19.9	1154	18.6
1400	16.9	1340	12.7
1600	14.0	1666	8.7

about the same comonomer ratio as 830 EW Dow polymer, but still a higher water content. One could explain that the longer branch of the Nafion ionomer increases water uptake by decreasing crystallinity. If water content is expressed as % weight gain, then these two polymers are comparable.

A similar effect is observed when comparing Nafion with the 3M polymer. Both the Dow and the 3M monomer lack the pending trifluoro methyl group and the second ether linkage of the Nafion monomer (see Table 3.1). In addition, the Dow monomer is shorter by two carbon atoms. The MW of the 3M monomer is 66 units lower than that of Nafion. Because the true EW of Nafion is about 30 units lower than the nominal value, a nominal 1100 EW has about the same comonomer ratio as 1000 EW 3M polymer. The water uptake of the two polymers is shown in Fig. 4.10. Note that in this plot the water content is shown per ion exchange site. Based on polymer weight, the 3M polymer would absorb about 10% more water in comparison with the 1100 EW PFSA; that is, the water uptake of the two polymers would be more similar. On the other hand, a solution cast film will usually exhibit a higher water uptake than a film based on extruded

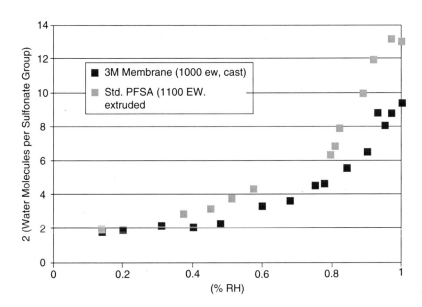

Figure 4.10 Moles of water absorbed per exchange site for Nafion and 3M polymer [10].

precursor polymer. Inherently, the 1000 EW 3M polymer therefore appears to have a somewhat lower water absorption than the 1100 EW PFSA.

The water uptake of Hyflon and Nafion as a function of EW are compared in Fig. 4.11. Again, for a given EW, the shorter branch polymer exhibits lower water absorption.

Figure 4.12 compares the water uptake of Hyflon-Ion with that of the earlier Dow polymer. Even though both polymers are made from the same monomers, there are some differences: The data for the Dow polymer appear to be more scattered and with lower water uptake at an EW below 850, possibly due to problems of achieving reproducible MWs and uniform distribution of both MW and EW.

The nature of the counter-ion is also an important factor in determining solvent uptake. It is the largest for the hydrogen ion form of the ionomer, particularly for solvents capable of forming an oxonium ion. For other ions, the water uptake decreases with decreasing hydration of the cation, for the alkali metal ions this means that with increasing atomic weight. The water uptake is particularly low for large organic cations, such as tetra butyl ammonium.

The importance of the thermal history on the swelling characteristic of perfluorinated ionomers has been demonstrated by Hinatsu et al. [12]. Each of six samples of polymer (two each Aciplex, Flemion and Nafion) ranging in EW from 890 to 1200 were pretreated by boiling in water for

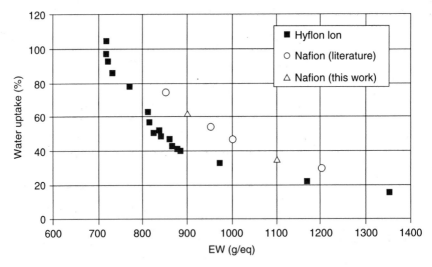

Figure 4.11 Water uptake of Hyflon-Ion and Nafion [5].

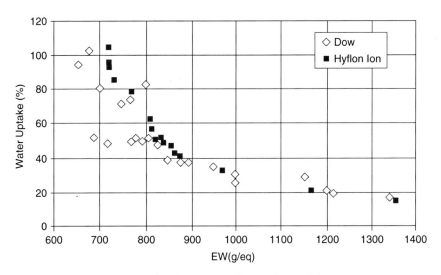

Figure 4.12 Water uptake of Hyflon-Ion vs. Dow polymer [5].

2 h (designated E), vacuum drying for 3 h at 80°C (designated N) and vacuum drying at 105°C for 3 h (designated S). (Drying times of 20 min were actually sufficient to give the observed pretreatment effect.) The samples were then immersed in liquid water at temperatures ranging from 25°C to 140°C and the water uptake determined (Table 4.8). The samples pretreated by drying swelled in water at the lower temperature substantially less than the boiled samples; however, at immersion temperatures above 110°C this effect of drying was erased. In general, one could say that the effect of drying at a given temperature is erased by immersion in water at a temperature slightly higher than the drying temperature and, conversely, the effect of immersion in hot water is erased by drying at a temperature slightly higher than the immersion temperature.

The water uptake from saturated water vapor at 80°C was substantially less than from liquid water at that temperature; however, the samples may not have reached equilibrium.

While Hinatsu described how the water uptake is reduced on drying at elevated temperatures, Coalson and Grot [13] reported how the water content is increased by exposing the ionomer to liquid water at temperatures above 100°C in an autoclave. Table 4.9 shows the water uptake of 1365 EW Nafion after pre-swelling at 100°C and 180°C followed in some cases by drying at ambient conditions or at 100°C under vacuum. The water

Table 4.8 Water Uptake from Liquid Water at 25°C after Various Pretreatments (in % based on dry polymer)

	N 125	N 117	AC 12	AC 4	FL 12
EW	1200	1100	1080	980	920
Boil (E)	22.5	37	40	44	64
Dry (N)	15	22	23	23	33
Dry (S)	13	18	18	20	26

Table 4.9 Water Uptake as a Function of Swelling and Drying Conditions for 1365 EW Nafion (in % Based on Dry Polymer, Autogenous Pressure)

Pre-swelling	None	100°C	100°C	100°C	180°C	180°C
Drying	None	None	20	100°C vacuum	None	100°C vacuum
RT swelling	16 %	26 %	19 %	14 %	98 %	
100 swelling	25 %	26 %	26 %	25 %	98 %	63 %

Table 4.10 Water Uptake of 1250 EW Nafion at Autogenous and Super-Autogenous Pressure

Temperature	100°C	150°C	150°C	175°C	200°C	200°C	225°C
Pressure	Ambient	Autogenous	35 atm.	70 atm.	352 atm.	703 atm.	703 atm.
% water uptake	28	39	86	141	279	370	518

Source: Ref. [13].

uptake is then measured at room temperature and after 10 min of boiling in water.

It can be seen that drying under vacuum at 100°C completely erases the effect of prior boiling in water, but only partially the effect of heating in liquid water to 180°C. In Table 4.6 the exposure to 180°C liquid water was done under autogenous pressure; even higher degrees of swelling can be obtained at increased pressures using an inert gas (Table 4.10). This may be related to the fact that on swelling, the increase in volume of Nafion is less than the volume of the water absorbed.

The need for an autoclave can be eliminated by using a higher boiling solvent, such as ethylene glycol or diethylene glycol. The higher boiling solvent is later leached out and replaced with water: For example, a 1200 EW film of Nafion was heated in ethylene glycol for 10 min to 184°C. The sample was cooled and the ethylene glycol leached out with water. At this point the water uptake was 209% based on dry polymer. The sample was then dried for 30 min at 100°C under vacuum. The water uptake of the dried sample was 95% in boiling water, compared to 28% for a sample of the same polymer not heated in ethylene glycol.

4.2.2 Wetting and Contact Angle

When a drop of water is placed on a dry film of Nafion, the film for the first few minutes appears fairly hydrophobic. This is particularly true for the samples that had been annealed for 30 min at 130°C in an atmosphere of dry nitrogen. Only when the film absorbs some water, the hydrophilic properties of the polymer become apparent. The contact angle of Nafion 112 as a function of time is shown in Fig. 4.13. The curves for the two annealed films (cast and extruded) are identical and show the least reduction of contact angle with time. This is a further indication that annealing

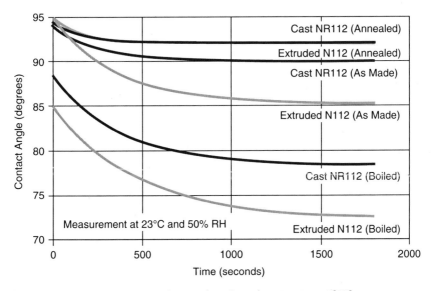

Figure 4.13 Water contact angle as a function of pretreatment [18].

erases the past history of the sample. Because film casting involves some annealing in the drying step, the "as made" cast film also has a higher contact angle than its extruded counterpart. Boiling the samples results in a substantial reduction of the contact angle [18].

The state of water imbibed by ionomers has been studied by DSC [21,22]. Additional information has been presented more recently at the DOE Hydrogen Program, May 2005, in Washington, DC. In Fig. 4.14a and b the results that Kim *et al.* obtained for Nafion of different water contents are shown (paper FC41). It appears that for water contents of less than about 12% by weight or about 7 mol per proton, none of the water freezes at the normal freezing point.

Newman (Fig. 4.15) reported a DSC study in which Nafion equilibrated in liquid water was first cooled at a rate of 10°C/min, later at 1°C/min. He observed two freezing points (−16°C and −27°C) representing different states of water. The sample was then heated at a rate of 10°C/min indicating melting points of water at −3°C and +0.4°C.

4.2.3   Solubility of Gases in Fluorinated Ionomers

Ogumi *et al.* [23] reported the solubility of hydrogen in Nafion of 1200 EW as 11.8 mmol/l for the Li form, 13.6 for the Na form and 9.7 for the K form. The solubility for oxygen is 10.9 mmol/l for 1200 EW and 10.7 for

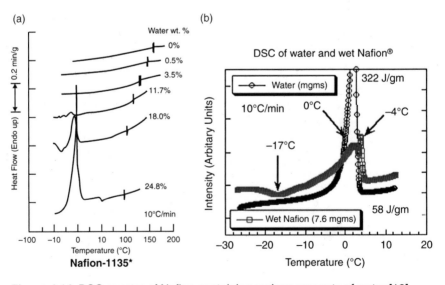

Figure 4.14 DSC spectra of Nafion containing various amounts of water [19].

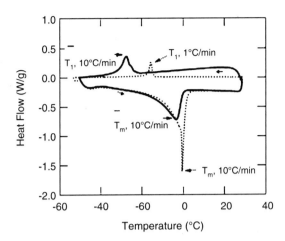

Figure 4.15 DSC of water in Nafion at different scan rates [20].

1100 EW Nafion, both in the K form. For comparison, the solubility of hydrogen and oxygen in water are 0.78 and 1.3, in perfluoro heptane 6.8 and 26.8 mmol/l and in pTFE 31.3 and 26.8 mmol/l, respectively. Based on the cluster model of Nafion, these results may indicate that the solubilities of these gases in the ionic clusters is similar to that in water, while the solubilities in the fluorocarbon matrix is similar to that in pTFE.

Tsou *et al.* [24] reported much lower values for the solubility of hydrogen in the Dow polymer: For polymer in the Na form, the solubility ranged from 0.8 mmol/l for EWs of 1030–1150 to 1.85 mmol/l for 840 EW and 2.97 mmol/l for 770 EW.

For the application in fuel cells, the solubility of gases in the hydrogen form of the ionomer is of most interest. Yeo and McBreen [25] reported a value 24.5 mmol/l for the solubility of hydrogen in 1200 EW Nafion in the hydrogen form.

The above results may indicate that compared to Nafion in the alkali ion form, the solubility of hydrogen gas in Nafion in the hydrogen ion form is substantially higher, and in the Dow polymer it is substantially lower. However, it should be noted that these results have been reported by three different teams with possibly differences in their procedures.

4.3 Morphology

Because Nafion has been readily available for many years, its morphology as an extruded shape, a liquid composition or as a film cast from the liquid composition, has been the subject of extensive studies. The results of these have not always been consistent. As Mauritz put it in a very recent review [16]: "... The quest for a universally accepted model continues with a spirited debate in the current literature."

One reason for this may be that the morphology of Nafion products can change: The morphology of ionomers, such as Nafion 117, derived from extruded precursor polymer will be changed on heating in liquid water, particularly to temperatures above the atmospheric boiling point (i.e., in an autoclave). These changes will be retained even after the material has been dried at room temperature [13]. Conversely, the morphology of a film obtained by the room temperature evaporation of a liquid composition of a fluoro ionomer is entirely different from that of a film obtained by the hydrolysis of precursor polymer and can be changed by dry heating or "curing" of this film.

Regardless of the disagreement about the details, the morphology of Nafion derived directly by the hydrolysis of the precursor polymer can be described in general terms as a network of ionic clusters imbedded in a continuous phase of fluorocarbon polymer. The ionic clusters contain the fixed ionic groups (sulfonate or carboxylate), their counter-ions and water of hydration. The fluoropolymer phase consists of backbone chain and part of the branches connecting the fixed ionic groups to the backbone. There is general agreement that there is a degree of crystallinity in the fluoropolymer phase. The clusters are interconnected by narrower channels, which control the transport properties of the polymer, while the equilibrium properties are determined by the clusters. An ionomer film in equilibrium with a solution of sodium hydroxide, for instance, may contain a significant amount of sodium hydroxide (as determined by titration), but pass hydroxide ions only at a very slow rate (see also Section 8.4). The size of the clusters and channels is obviously a function of the degree of hydration.

To illustrate the proportions of the various phases the example used in Section 8.4 may be useful: 1046 g (dry) of Nafion of 1046 EW may imbibe 180 g of water (=10 mol per exchange site) under conditions of partial hydration. This polymer would contain 681 g of fluoropolymer backbone (six TFE plus one trifluoro vinyl group), 282 g "tethers" (a tether is a branch minus the ionic group, it ties the ionic group to the backbone) consisting of four carbons, two ethers and one pendent trifluoro methyl groups and 260–300 g of ionic phase consisting of water, fixed ions, counter-ions and possibly ions of the external electrolyte, which have overcome the Donnan repulsion and invaded the cluster. It is interesting to note that dimethyl methylphosphonate is a solvent that is reported to swell only the tether phase of Nafion [10]. The volume relationship between these three phases is approximately 310 cm^3 for the fluoro polymer phase, 140 cm^3 for the tether phase and about 260 cm^3 (=37%) for the ionic phase. The volume of the ionic phase may increase to 440 cm^3 (=49%) for higher degrees of hydration (20 mol per exchange site).

It is apparent that

1) The ionic clusters must be surrounded by a shell of tethers. This shell may also contain a few ionic groups, whose tethers are attached to backbones further away from the cluster.

2) If all three dimensions of the cluster are much larger than twice the length of the tether, the interior of the cluster must contain a core free of fixed ions.

This condition would exist, if the water or solvent content of the ionomer is substantially increased, for instance by heating the ionomer in water or a water/solvent mixture in an autoclave to temperatures in the 130–200°C range. In this state, the ionomer would contain large clusters mostly filled with water or solvent and only the walls of the clusters would be lined with a layer of fixed ionic groups surrounded by a tether shell. Such an inverted micelle could be viewed as a transition to the liquid composition.

The original model proposed by Gierke [15] incorporated spherical clusters of about 4 nm diameter interconnected by cylindrical channels of about 1 nm diameter. The large diameter of the cluster would allow an interior core relatively free of fixed ionic charges, and, therefore, less subject to Donnan repulsion of anions. This would explain the amount of anions entering the ionomer in *equilibrium* with an external electrolyte solution. The much smaller diameter of the channels would give the electrostatic repulsion of anions responsible for the limited *rate* of anion transport. An extensive review, including 243 references, of the morphology of solid Nafion has been published recently [16].

It is understood that the spherical clusters and cylindrical channels of the Gierke's model are idealized geometries, selected to facilitate mathematical treatment. Some irregularities must be assumed. However, other researchers have suggested entirely different geometries: Haubold [26] proposed a flat, sandwich-like structure and Gebel [27,28] a rod-shaped geometry.

4.3.1 Morphology of Liquid Compositions

The morphology of the liquid compositions depends very much on the conditions during dissolution. Two solutions made from the same polymer and with the same solvent composition may have different properties, such as viscosity and film-forming ability. Furthermore, these properties may be changed by a posttreatment of a given solution (see below Curtin *et al.*). Each of the various suppliers uses their own proprietary procedures to achieve a certain composition, except that the liquid compositions offered by Aldrich Chemicals and Solution Technology appear to be identical. It is, therefore, not surprising that researchers have reached different conclusions about the morphology of these liquid compositions.

In general terms, the morphology can be described as dispersed ionomer particles (micelles) consisting of a fluoropolymer core surrounded by ionic groups extending into the solvent phase. Some counter-ions will diffuse into the solvent phase in a dynamic equilibrium (Donnan equilibrium), leaving the micelle with a negative charge, which stabilizes the micelles by electrostatic repulsion. These micelles may consist of one or more individual polymer molecules.

Lee studied the properties of dilute (0.2–9.0 mg/ml) solutions of Nafion in 80% methanol using a membrane osmometer, viscoelastic analyzer and dynamic light scattering. Different size aggregates were observed depending on the concentration: Below 1 mg/ml, rod-like primary aggregates of about 1000 nm size were observed. They were attributed to the hydrophobic interaction of the fluoropolymer backbone. Above 1 mg/ml, larger aggregates (10,000 nm) were observed believed to be due to electrostatic dipole/dipole attraction of the primary aggregates. Dissolution in propanol/water mixtures breaks these aggregates down to individual polymer molecules [17]. Curtin *et al.* determined the MW distribution of aqueous dispersions of Nafion by size exclusion chromatography and found a bimodal distribution with a main peak at about 5×10^5 g/mol and a smaller one at about 4×10^6 g/mol due to aggregates. On heating the aqueous dispersion to 230°C, 250°C and 270°C, these peaks gradually disappear, suggesting a breakup of the aggregates. A newly formed peak at about 1.7×10^5 g/mol had a shape close to that of a Gaussian distribution. The value of 1.7×10^5 g/mol is in good agreement with measurements of the MW of both Nafion and Flemion. It appears therefore that on heating in water to 270°C the micelles are broken down to individual polymer molecules, and that the aggregates represented by the initial peaks contained about 3 and 20 polymer molecules, respectively. Light scattering revealed that the radius of gyration was a linear function of the molar mass of the aggregates, suggesting some elongated structure of the micelles (Fig. 4.16) [18].

The morphology of a film cast from a liquid composition is dependant on the temperature of solvent removal and on any subsequent annealing or

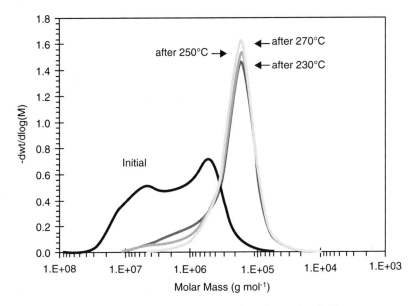

Figure 4.16 Mass of Nafion micelles after heating dispersion [18].

"curing" step: If the solvent is evaporated at room temperature, then the resulting film largely retains the morphology of the liquid composition from which it was cast. Such a film can be readily re-dissolved in methanol at room temperature, because it has not formed the continuous fluoropolymer phase discussed above. This could be described by the "bricks and mortar" model: Even though the fluoropolymer phase (the "bricks") occupies a large volume fraction, it is discontinuous. If such a film is heated to 100–150°C, this methanol solubility is greatly reduced and its equilibrium water absorption approaches values for that of films made from extruded precursor polymer. Compare also with Section 4.2.2. Siroma reports that heating a cast film for 1 h to 120°C or 1 min to 160°C greatly improves the room temperature solvent stability of the cast film. However, even the cured film exhibits greater swelling in hot methanol than Nafion 117 [11].

4.4 Transport Properties

Almost all applications of fluorinated ionomers involve the desired transport of some species and the prevention or minimization of the transport of any undesired species. In most cases, such transport is driven by an electrical current or a potential difference across the membrane. In some cases, the transport may be driven by a difference in concentration of some dissolved species or a difference of partial pressure of a gas or vapor. In addition, even in cases where the electrical forces are the dominant driving force, concentration difference between the two electrolytes could be a factor in the transport rates.

Pressure difference is a minor factor in determining transport phenomena. While Nafion can be used as a reverse osmosis membrane, rates are slow and an application that would justify the price of a fluorinated ionomer has not been found.

4.4.1   Transport Driven by an Electric Current

This is by far the most important application of fluorinated ionomers. It is discussed in detail in the chapter on applications (Chapter 5).

4.4.2   Transport Driven by a Concentration Difference

These are discussed in Sections 5.3 and 5.4.

4.5   Optical Properties

Unreinforced fluorinated ionomers are clear films with no absorption in the visible spectrum, except for those introduced by colored cations. The refractive index of Nafion in the dry hydrogen ion form is about 1.35; water uptake reduces it to about 1.34. This value is very close to the refractive index of water. As a result, highly swollen Nafion is almost invisible in water, a property that has been utilized to study the growth of microbial films with an "invisible" support.

The typical perfluorinated ionomer films, such as Nafion, are made in a continuous process where the sheet is always under tension in the MD. As a result, there is a degree of orientation introduced into the film that results in birefringence, as well as differences in mechanical and swelling properties in the MD and TD. This orientation is already present in the extruded precursor film. This birefringence becomes apparent if a sample of the films is examined between crossed polarizers with the MD at a 45° angle against the plane of polarization. The sample will light up against the dark background. Stretching in the MD will further lighten the sample, while gentle stretching in the TD will first darken it, followed by lightening, as the original orientation is first cancelled and then reversed.

In reinforced membranes, the birefringence is greatly reduced or absent. It appears that the heat applied in lamination relaxes the orientation of the precursor film and any new stresses on the membrane are picked up by the reinforcement.

4.6 Thermal Properties

The glass transition temperature of 1100 EW Nafion in the hydrogen ion form has been reported as 105°C. Ikeda *et al.* [29] report glass transition temperatures of 123°C for Nafion, 148°C for the Dow polymer and 144°C for the 3M polymer The glass transition temperatures of the Li, Na and K form of Nafion are 215°C, 235°C and 225°C, respectively.

4.7 Stability

The excellent chemical stability of perfluorinated ionomers is apparent from many of its applications: They exhibit long-term stability to wet chlorine at 90°C either in pH 3 brine or in 20% HCl, 32% NaOH at 90°C, hot chromic acid and chromic/sulfuric acid mixtures, hot nitric acid up to 100% concentration, 70% nitric acid at 120°C, etc. There is some concern about the exposure to peroxide radicals; for instance, as they are generated in fuel cells (compare Section 6.3).

The effects of X-rays on Nafon 117 were studied by Almeida and Kawano [30]. The EPR spectrum of irradiated Nafion indicated the presence of peroxide radicals similar to that found in irradiated pTFE. IR spectra showed new bands at 1458, 1700, and 1773/cm, indicating the presence of C=O and C=C groups.

The thermal decomposition of fluorinated ionomers is to some extent a function of the ionic form. The acid form, particularly in the dry state, is less stable than other ionic forms. DuPont recommends 175°C as the maximum long-term use temperature for Nafion in the acid form. The presence of solvents that can be protonated, such as water, alcohols and ethers, stabilizes the acid form, presumably due to the formation of oxonium salts. For instance, liquid compositions of Nafion can be exposed to temperatures of 270°C and higher without any apparent adverse effects. Ikeda *et al.* report the decomposition temperature in air for 950 EW Nafion or Flemion and 740 EW Dow polymer as 319°C, but 362°C for the 3M

Table 4.11 Thermal Degradation Products of NAFION Sulfonic Acid
Polymer

Compound	Evolution Temperature °C	Mg/g Sample
SO_2	280	15
CO_2	300	30
HF	400	—*
CO	400	3
R_fCOF	400	10**
COF_2	400	3
COS	400	Trace
R_fOH	400	Trace

*Significant level but could not calculate because HF reacts with and absorbs on cell walls.
**Mixture of products.
Source: Ref. [31].

(800 EW) polymer. In argon, the values are 317°C for Nafion or Flemion, 312°C for Dow polymer and 393°C for 3M polymer [29]. It appears that the higher values for the 3M polymer are due to the extra two carbon atoms between the sulfonic acid group and the ether linkage. Furthermore, the degradation of the 3M polymer in air appears to be oxidative, while the degradation of the other polymers is purely thermal.

The thermal degradation products of Nafion in air are shown in Table 4.11. A 500-mg sample was heated in a stainless steel tube in an air flow of 13 ml/min. The temperature was raised initially at 10°C/min to 200°C, then at 5°C/min to 400°C and held there for an additional 20 min: The amount of sulfur released as sulfur dioxide represents about 25% of the sulfur content of the sample. It is not clear whether additional sulfur dioxide would be released at higher temperatures.

References

1. Gilbert, M., Haworth, B., Myers, D., Polym. Eng. Sci., **44**(2), 272–282. 2004.
2. Gierke, T., Munn, G., Wilson, F., J. Polym. Sci. Polym. Phys. Ed., **19**, 1687–1981.
3. Starkweather Jr., H., Macromolecules. **15**, 320–1982.
4. Hsu, W., Gierke, T., J. Memb. Sci., **13**, 307–1983.

5. Ghielmi, A., Vaccarono, P., Troglia, C., Arcella, V., J. Power Sourc., **145**, 2108–115, 2005.
6. Doyle, M., Lewittes, M.E., Roelofs, M.G., Perusich, S.A., J. Phys. Chem. B., **105**, 9387–9394, 2001.
7. Gebel, G., Aldebert, P., Pineri, M., Polymer. **34**(2). 333–339, 1993.
8. Grot, W., Munn, G., Walmsley, P., Paper presented at the 141th ECS Meeting, Houston, 1972.
9. Arcella, V., Ghielmi, A., Tommasi, G., Ann. N.Y. Acad. Sci., **984**, 226–244, 2003.
10. Debe, M.K., 2005 DOE Hydrogen Program Review FC3, Washington, DC, May 2005.
11. Siroma, Z., Fujiwara, N., Ioroi, T., Yamazaki, S., Yasuda, K., Myazaki, Y., J. Power Sourc., **126**(1-24), 1–45, Feb 2004.
12. Hinatsu, J., Mizuhata, M., Takenaka, H., J. Electrochem. Soc., **141**(6), 1493–1498, 1994.
13. Coalson, R., Grot, W.G., US Patent. 3,684,747 assigned to DuPont, Aug. 15 1972.
14. Schneider, N.S., Rivin, D., Polymer. **45**(18), 6309–6320, 2004.
15. Gierke, T.D., Hsu, W.Y., J. Memb. Sci., **13**, 307, 1983.
16. Mauritz, K.A., Moore, R.B., Chem. Rev., **104** 10, 4535–4585. 2004.
17. Lee, S., Yu, T., Lin, H., Liu, W., Lai, C., Polymer. **45**(8), 2853–2862, April 2004.
18. Curtin, D., Lousenberg, R., Henry, T., Tangeman, P., Tisak, M., J. Power Sourc., **131**(1-2), 41–48, 2004.
19. Kim, Y.S., 2005 DOE Hydrogen Program Review FC41, Washington, DC, May 2005.
20. Newman, J., 2005 DOE Hydrogen Program Review FC50, Washington DC. May 2005.
21. Yoshida, H., Miura, Y., J. Memb. Sci., **68**(1-2), 1-10. April 10, 1992.
22. Kim, Y.S., Macromolecules. 36, 176181–2003.
23. Ogumi, Z., Kuroe, T., Takehara, Z., J. Electrochem. Soc., **132**, 2601–2605, 1985.
24. Tsou, Y., Kimble, M.C., White, R.E., J. Electrochem. Soc., **139**(7), 1913–1917, July1992.
25. Yeo, R.S., McBreen, J., J. Electrochem. Soc., **126**, 1682–1979.
26. Haubold, H.-G., Jungbluth, Th., Hiller P. Electr.chim. Acta. **46**, 1559–1563, 2001.
27. Gebel, G., Rubatat, L., Diat, O., Macromolecules. **37**, 7772–7783, 2004.
28. Gebel, G., Rubatat, L., Diat, O., Rollet, A., Macomolecules. **35**, 4050–4055. 2002.
29. Ikeda, M., Uetmatsu, N., Saitou, H., Hoshi, N., Hattori, M., Iijima, H., Polym. Preprint., Japan. 54 2, 2004.
30. Almeida, S.H., Kawano, Y., Polym. Degrad. Stabil., **62**(2), 291–297. Nov 1998.
31. DuPont NAFION Bulletin E-63118-1, Safety in Handling and Use. June 2000.

5 Applications

While the applications for most fluorinated polymers, as well as other polymers, merely require stability in a sometimes hostile environment in terms of corrosive chemicals, high temperatures, strong mechanical stresses, or electric fields without failure, the applications of fluorinated ionomers are based on their ability to actively engage the surrounding environment: as catalysts they cause chemical reactions, as membranes they allow the flow of an electric current under small electric fields (the smaller the better). They can discriminate between anions and cations or allow the passage or water while acting as a barrier for gases. As an actuator (or artificial muscle), they create mechanical stresses or movement as a result of an electrical stimulus.

From a commercial standpoint, the most important examples of such interaction with a chemical system are found in the use of fluorinated ionomers in membrane form as separators in electrolytic cells.

5.1 Electrolysis

5.1.1 Introduction, Electrolytic Cells

In an electrolytic cell, an electric current is passed from an electronic conductor through one or several ionic conductors (electrolytes) back into a second electronic conductor. The circuit is closed outside of the cell through various electronic conductors, typically including a power supply and a current measuring device. The junctions between the electronic and ionic conductors are called electrodes, namely cathodes and anodes depending on the direction of current flow: at the cathode, electrons are transferred from the electronic conductor to some component of the electrolyte causing the discharge of some positive ion (cation) or the generation of a negative ion and also the reduction of some component of the electrolyte. As a common example, the transfer of an electron from the cathode to a water molecule causes the formation of an OH^- ion and a hydrogen atom. Two of these hydrogen atoms will then combine to form a hydrogen molecule. The opposite happens at the anode: in a typical case, a chloride ion transfers its extra electron to the anode and is converted to a chlorine atom. It should be noted that the terms "anode" and "cathode" indicate the direction of current flow and not the polarity: in a cell that consumes electric power, such as in NaCl electrolysis, the anode is positive; and in a cell that generates electric power, such as a fuel cell, it is negative. In a rechargeable battery, the electrodes switch function between charge and discharge.

In some cases, an undivided cell can be used in which both anode and cathode share a common electrolyte. The application for ionomer membranes is of course in cases where a divided cell is desirable or even necessary to prevent interference of the anodic oxidation with the cathodic reduction and allow separate collection of the reaction products.

The most common type of divided electrolytic cell uses the "plate and frame" design, in which up to 160 individual cells are stacked up in a unit called "electrolyzer" or "stack". This cell design will be discussed in more detail in Section 5.1.2.

In cases where the electrolysis involves the addition or removal of solids, an open tank type cell is more practical. In this cell design, an open tank is filled with the "working" electrolyte and one or several "working" electrodes (i.e., anodes, in the case of the anodic dissolution of a metal) and "counter-electrodes" are suspended in this tank from the top. The counter-electrodes are placed inside a bag made of ionomer membrane and filled with counter-electrolyte. In this way, the addition or the removal of solids on the electrodes is facilitated.

In dealing with particularly hazardous or corrosive electrolytes, a closed tank type cell is sometimes used. Such a cell is often cylindrical with minimum external gasketing. An example of such a cylindrical cell is shown in Section 5.1.8.

5.1.2 NaCl Electrolysis

This is by far the largest application for fluorinated ionomers. The process is illustrated in Fig. 5.1.

The main anode reaction is represented by $2NaCl \rightarrow 2Na^+ + 2e^- + Cl_2$. The sodium ions move through the membrane into the cathode compartment while the electrones, pushed by the power supply, travel through the external circuit to the cathode. There they react: $2e^- + 2Na^+ + 2H_2O \rightarrow 2NaOH + H_2$. The current flowing through any cross-section of the cell is equal to the sum of negative charges moving to the left and positive charges moving to the right. The distribution between these two modes of transport is quite different in the three ionic conductors in the cell: in the membrane, the transport is almost exclusively by positively charged sodium ions; in the catholyte, it is predominantly by hydroxide ions;

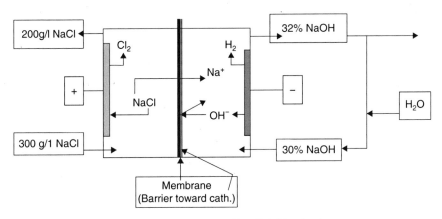

Figure 5.1 Schematic of the electrolysis of NaCl (side view).

Table 5.1 Equivalent Conductance of Some Ions at Four Different Temperatures

Ion	0°C	25°C	75°C	100°C
H⁺	240	350	565	644
Li⁺		39		
Na⁺	26	51	116	155
K⁺	40	75	159	206
OH⁻	105	192	360	439
Cl⁻	41	75	160	207

Note: Values are in ml/equiv. $\times \Omega \times$cm, extrapolated to infinite dilution.

and in the anolyte, more than half by chloride ions (in both the cases the balance is by sodium ions). The reason for this can be seen in Table 5.1.

The value for the hydrogen ion, which is not part of this system, is given only for comparison. While the values shown are for infinite dilution, the same relative order applies to more concentrated solutions. It can be seen that the OH ion is by far the most mobile one of the ions involved. The prevention of its movement through the membrane was the most challenging task in the development of the membrane technology for the chlor-alkali process. It was made more difficult by the fact that the movement of this ion through the membrane is driven by both a concentration gradient and the difference in electrical potential. Any OH ion that migrates into the anode compartment represent a failure to transport a sodium ion, and therefore a loss in current efficiency (CE). It also requires addition of HCl to the anolyte to prevent the formation of sodium hypochlorite.

The transport of chloride ion through the membrane is driven by the concentration gradient, but retarded by the difference in electrical potential. The resulting contamination of the catholyte, therefore, decreases with increasing current density (CD).

As a result of the change in current carriers at the four phase boundaries in the cell, there are concentration changes at these boundaries: depletion of NaCl at both the anode/anolyte and the anolyte/membrane interphase. The formation of NaOH in the catholyte occurs primarily at the membrane/catholyte interphase and only to a small extent at the surface of the cathode.

All industrial scale chlor-alkali plants incorporate the "plate and frame" design, in which many membranes and electrodes are assembled together with frames and gaskets in an electrolyzer. The electrical connection between the electrodes can be either in parallel (monopolar) or in series (bipolar) as illustrated in Fig. 5.2. In a monopolar electrolyzer, all anodes are connected to the same positive bus bar and all cathodes are similarly connected to the same negative one. As a result, the voltage for the entire electrolyzer is equal to the individual cell voltage (about 3 V). Also, all cells in a single electrolyzer operate at the same voltage and any difference in their internal resistance will be reflected in an uneven current distribution.

In contrast, in a bipolar electrolyzer, the only electrical connection is made at the two single-faced electrodes at the ends of the stack; the bipolar electrodes are without electrical connections. The electrolyzer voltage is equal to the sum of the voltages of all the individual cells (typically 200–400 V). All cells carry the same current and any differences in their internal resistance will be reflected in differences in cell voltage, and therefore easily detectable.

The DC power supplies used in the process operate most desirable at a DC voltage of 200–400 V. For a given production, a lower voltage would require a higher current and therefore more massive conductors. A higher voltage would be of concern from a safety standpoint because this voltage is not only carried by the electrical conductors and cell components, but also by the electrolytes and associated equipments such as pumps and valves. One advantage of a bipolar electrolyzer is, therefore, that it can be directly connected to a power supply. Plants that employ several bipolar electrolyzers, which is usually the case, will frequently feed each electrolyzer from

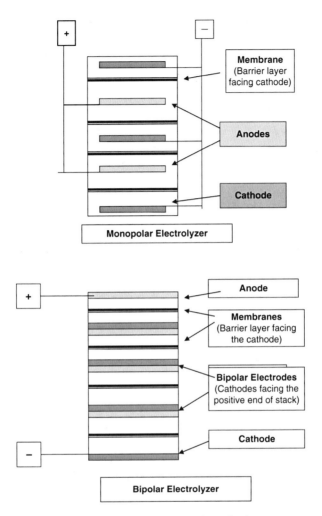

Figure 5.2 Monopolar and bipolar electrolyzers (top view).

a dedicated power supply, which allows for greater flexibility in terms of electrolyzer maintenance and adjusting production rates.

Monopolar electrolyzers on the other hand are almost always used in a series connection of many electrolyzers to match the optimum output voltage of the DC power supply. The individual monopolar electrolyzers are therefore usually of smaller capacity than their bipolar counterparts.

One problem unique to bipolar electrolyzers is "bypass" currents flowing through the electrolyte manifolds. As the electrolytes are fed to all the individual cells of an electrolyzer from common manifolds, the substantial

Table 5.2 Commercially Available Electrolyzers

Manufacturer	Type	Polar	Active area (m^2)	Max. CD (kA/m^2)	Membrane dimensions (cm)	Electrode gap (mm)
Uhde	BM2.7 v 3	Bi	2.7	6	122 × 240	1.2
Uhde	BM2.7 v 4	Bi	2.7	6		0.4
ElectroCell	ElectroProd	Mono or Bi	0.4	4	63 × 85	1–4
ElectroCell	ElectroSyn	Mono or Bi	0.04	4	18 × 38	1–5
ElectroCell	MP	Mono or Bi	0.01	4	13.8 × 18.5	1–12
Asahi Kasei	Super	Bi	5.1	4		
Asahi Kasei	Standard	Bi	2.7	4		
Chlorine	CME	Mono	3.03	4		
Engineers	BiTAC	Bi		6		
ICI	FM 21	Mono	0.21	4		

potential differences between the cells cause a fraction of the current to flow through the manifolds, thereby bypassing the stack. This not only represents a loss in production, but may also cause corrosion problems within the cells. During a power interruption, chlorine and other reactive components cause the cells to act like batteries, which discharge through the electrolyte manifolds. This represents a current reversal within the cells. Proper cell design can minimize or eliminate these problems.

Commercially available electrolyzers are shown in Table 5.2.

Two designs are selected for a more detailed discussion: the Uhde–DeNora Cell (Fig. 5.3), as an example of a large bipolar electrolyzer in worldwide use for the electrolysis of sodium chloride, and two different sizes of the ElectroCell designs, as examples of smaller monopolar cells suitable as general-purpose cells.

5.1.2.1 *Uhde Cell*

The Uhde cell was selected because of its worldwide acceptance as a high-performance electrolyzer with increasing market share (compare Fig. 5.5). Its design is unique in several respects: it is the only electrolyzer in

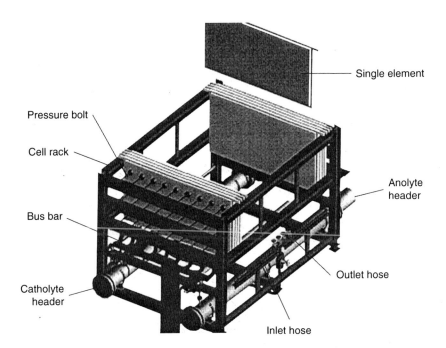

Figure 5.3 The Uhde cell stack [82].

which the single cells comprising the electrolyzer are individually sealed. It is therefore possible to remove one cell from the electrolyzer without disturbing the sealing of the entire stack.

In all the other electrolyzers, both mono- and bipolar, the entire stack is sealed by forces applied to the two ends of the stack, usually by a single hydraulic ram working against one end of the stack, except the Asahi Kasei Super Stack, which employs two rams side by side. Replacing a single membrane in the stack, therefore, requires shutdown and opening of the entire stack.

Figure 5.3 shows a schematic of a partially assembled stack. A single element consisting of a sealed electrolytic cell is lowered into place. Up to 150 cells may be assembled in a single stack. The two electrolyte inlets and two outlets are then connected by hoses to the respective headers. Current is supplied to the stack through bus bars located at the two opposite ends of the stack.

Figure 5.4 shows the cell room of a fairly large chlor-alkali plant. About 10 electrolyzers or stacks are visible in the left side of the cell room; more on the right hand side.

The Uhde design is not strictly bipolar, because it lacks the bipolar electrodes consisting of an anode and a cathode attached to each other as a single component. Instead, the individual cells (in Fig. 5.3 called "single element") consist of anode and cathode half shells, bolted and sealed

Figure 5.4 Cell room of a 1000 t/day chlorine plant [82].

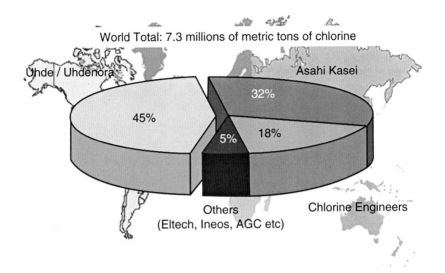

Figure 5.5 Market shares of some technology sellers [83].

together with the membrane in between. In the stack, the anode side of such a cell makes electrical contact with the cathode side of the adjacent cell. If a single membrane or other cell component needs to be replaced, that particular cell is lifted out of the stack and a preassembled replacement cell is inserted in its place.

Another interesting feature of the Uhde design is that both inlet and exit headers are *under* the electrolyzer. In most other electrolyzers, the two exit headers are located above the electrolyzer. In the Uhde design, a plastic discharge pipe inside the half shells carries the exit electrolyte/gas mixture out through the bottom of the cell. The electrolyte volume inside the cell is relatively large: 110 l anolyte and 80 l catholyte. The large volume improves gas/liquid disengagement and, on the anolyte side, provides good internal mixing of the low pH, high concentration anolyte feed, with the bulk anolyte.

Other feature inside the cell are more or less common with other high-performance chlor-alkali cells: because cell voltage is a major factor determining operating costs, close attention is given to the gap between anode and cathode (Fig. 5.12). In the most recent design version 4, the electrode gap has been reduced to 0.4 mm. A slightly positive pressure on the cathode side is used to lean the membrane gently against the anode. Efficient liquid/gas disengagement is important on both sides. Gas generated between the electrode and the membrane must be quickly redirected to the larger space behind the electrode. On the anode side, a louvered electrode is used for this purpose. Also on the anode side, a baffle and a downcomer plate in the space behind the anode are used to create internal anolyte circulation powered by the gas lift – the gas dispersed in the electrolyte between the anode and the baffle creates a lift. The gas disengages at the top of the cell and the mostly gas-free liquid flows down behind the baffle to create effective mixing with more concentrated inlet anolyte (300 g/l inlet versus 200 g/l outlet).

5.1.2.2 ElectroCells

ElectroCell offers three different monopolar cell sizes; the largest (the ElectroProd cell) and the smallest (the MP cell) are selected for detailed discussion, because they are not only usable in chlor-alkali applications, but also as general purpose electrochemical cells. In addition to the typical operation with only a single membrane between the electrodes, they can also accommodate two or three membranes between the electrodes, dividing the cell into three or four compartments (Fig. 5.6). The extra compartments can also be used for the introduction of heating or cooling liquids. For their use as multipurpose cells, their injection molded frames are offered in a wide variety of inert materials, such as polypropylene or PVDF, and the cells can be equipped with different electrodes, including DSA, graphite led dioxide or platinum on titanium, as required.

The inlet and outlet headers are built into the cell frames, as it is common with mono polar cells. The outlet headers are on top. Closure is with tie rods and screws, or the larger ElectroProd cell can also be equipped with a hydraulic ram. By installing, in the center of the stack, a gasket that will block off all manifolds, it is possible to double the number of compartments between electrodes (eight compartments for the MP cell). An example of an application for such a stack is shown in Chapter 9 "Experimental Methods" under cation/cation equilibria (Fig. 9.2).

The ElectroProd cell has an active area of 0.4 m²/cell. Up to 40 cells can be stacked in a module for a total area of 16 m². The electrolyte flow rate per cell is 10–30 l/min resulting in linear velocities of 0.05–0.4 m/s.

Figure 5.7 shows the smaller MP cell. A fluid distributor made from a plastic mesh helps to provide uniform electrolyte distribution.

The MP cell has an active area of 0.01 m²/cell. Up to 20 cells can be stacked in a module for a total area of 0.2 m². The electrolyte flow rate per cell is 1–10 l/min resulting in linear velocities of 0.03–0.4 m/s. The electrolyte volume is 0.2 l/cell.

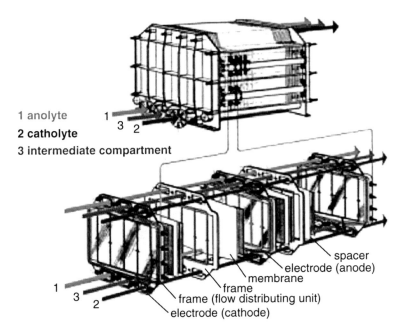

1 anolyte

2 catholyte

3 intermediate compartment

spacer
electrode (anode)
membrane
frame
frame (flow distributing unit)
electrode (cathode)

Figure 5.6 The ElectroProd cell [84]. Exploded view of three component monopolar module.

5.1.2.3 *Membranes*

For the industrial scale chlor-alkali electrolysis, the basic construction of all the membranes in use today is the same: an about 100–μm-thick support layer of perfluorinated sulfonic polymer, into which an open mesh reinforcing fabric made of polytetrafluoroethylene (pTFE) is imbedded. The three major membrane manufacturers (Dupont, Asahi Glass and Asahi Kasei) all use a copolymer based on the same molecular weight (MW) = 446 co-monomer first used in Nafion® for this sulfonic layer. The cathode side of the laminate is coated with a 20–μm-thick barrier layer made of a perfluorinated carboxylic copolymer. Asahi Glass uses a different monomer for this layer than the other two manufacturers (see Table 3.1). The cathode side or both sides of the membrane are then coated with a gas-release coating, consisting of finely divided zirconium oxide in a perfluorinated ionomer binder.

While some of the monomers used by the three companies are the same, the resulting copolymers may exhibit some differences in respect to MW and equivalent weight (EW), the distribution of these parameters and unstable end groups. The performance of the membranes can be judged in terms of cell voltage, perhaps the most important factor, CE, purity (low level of chloride) of the sodium hydroxide produced, sensitivity to

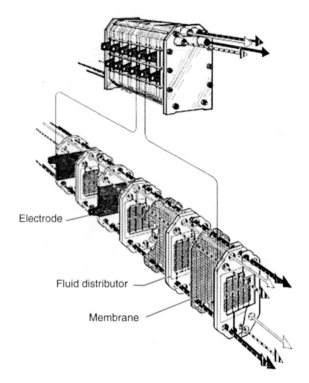

Electrode

Fluid distributor

Membrane

Figure 5.7 The smaller MP cell. Exploded view of the EC Electro MO cell.

brine impurities and operational upsets, mechanical strength and endurance. Because of the importance of cell voltage, membrane manufacturers have developed new "high performance" membranes. The cell voltage of four such "high performance" membranes as a function of "weeks on line" is shown in Fig. 5.8. The test conditions are: 3 kA/m^2, 32% NaOH, 90°C, using an Uhde electrolyzer equipped with third generation elements. The slight increase in cell voltage over a period of almost 2 years may be due to the accumulation of scale deposits. At the same time, technology sellers such as Krupp-Uhde are continuously trying to improve the cell design in an effort to lower cell voltage. Comparing the cell voltages of NAFION 981 WX and NAFION 982 WX (both "high performance" membranes) in Fig. 5.8 after 100 weeks of operation with that of NAFION 966 (a "standard performance" membrane) in an electrolyzer equipped with second generation elements, a voltage saving of about 20 mV can be seen, due to improvements in both cell design (the "generations") and membranes. NAFION 966 on the other hand is stronger and can better withstand mechanical stresses.

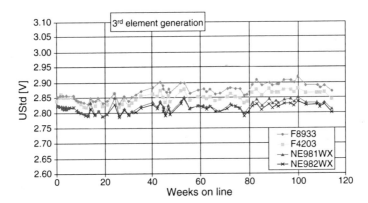

Figure 5.8 Cell voltage vs. time for some membranes [82]. NE981WX, NE981WX (Dupont); F4203 (Asahi Kasei); F8933 (Asahi Glass). Charts shows the standardized voltages (3 kA/m², NaOH 32% w/w, 90°C) of the membranes running parallel.

The technology sellers collect extensive data from the field to determine the most suitable membranes for equipping their cells. The current overall ranking of some membranes by one technology seller is shown in Fig. 5.9.

These membranes can produce a 30–35% solution of sodium hydroxide at high CE. However, they require very high-purity (less than 50 ppb Ca + Mg) sodium chloride as the starting material. For the generation of chlorine on a very small scale, the required purification of sodium chloride would not be practical. In such applications, for instance for swimming pool chlorinating or emergency water disinfecting, Nafion 324 can be used. In this membrane a high EW sulfonate barrier layer is used. While this polymer is less effective than the carboxylate polymer for rejecting hydroxyl ions, it is also much less sensitive to brine impurities. An 85% CE at 10–15% caustic concentration can be achieved, which is adequate for the applications mentioned.

5.1.2.4 Electrodes

Both electrodes are made of mesh or perforated material to allow gases formed between the electrode and the membrane to escape to the back of the electrode and to utilize the back surface of the electrode for the electrochemical reaction. Flattened expanded metal has been used or a special louvered design is used to, more effectively, remove gases from the electrode–membrane space, where they would increase the electrical

Membrane	Number	Wol Max.	U	CE	NaCl in NaoH	Resistance against Impurities	Mech. Stab.	Ranking
DuPont								
NX/N982WX	10700	240	+	+	+	+	+	1
NX2010WX	180 (30*)	80 (200*)	+	+/−	+		+	3
AGC								
F 893.2NT	40	240	+	+	+		+	4
F 893.3NT	430	240	−	+	+		++	3
F 893.5NT	620	170	+	+	+		+	2
F 8020	820	50	++	+	+			1
AKC								
F 4203	3580	240	+	+/−	−−	−	−−	6
F 4401	8	120	+	+	−−		−−	6
F 4401 C	590	110	+	+		−	−−	5
F 4402	30	130	+/−	+/−			−−	5

*NE2010WX

Figure 5.9 Membrane ranking [83].

resistance. The base material for such an electrode is a 1-mm-thick sheet of titanium. Small die cuts, horizontal in the picture, are made into this sheet. The die cut sheet is then stretched, vertical in the picture, which opens the cut to form hexagonal holes. At the same time, the ribs between the openings twist about 45°, increasing the thickness and exposing some sharp edges. The sheet is then flattened to about 1.5 mm. This is followed by coating with the catalyst layer.

Anodes: The older graphite electrodes have now almost completely been replaced by titanium based anodes. These "DSAs" (dimensionally stable anodes) were invented by Beer, and further improved and commercialized by Nora [1,2]. They are made by coating a titanium substrate with a catalytic layer consisting of a mixed oxide of a platinum group metal and titanium or another "valve" metal (Ta or Zr). For chlorine evolution, the platinum group metal is usually ruthenium in an amount of 20–45% by weight of the entire oxides. The overpotential for chlorine evolution is 90–120 mV at CDs of 2–10 kA/m^2. The overpotential for oxygen evolution is much higher, which is highly desirable for the chlor-alkali application; if oxygen evolution is desired, a coating containing iridium is chosen. An extensive review of these anodes can be found in Ref. [3] (Fig. 5.10).

These anodes can be damaged by fluoride ions, some organic acids such as formic or oxalic acid and by operation at a pH higher than 11.

Cathodes: The base metal is usually nickel. The hydrogen overpotential of nickel is about 300 mV. To reduce this further, "activated" coatings have been used. Some of these consist simply of high surface area nickel,

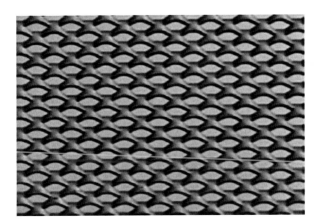

Figure 5.10 Anode made from expanded metal [82]. This figure shows the special structure of the anode mesh installed in DeNora based electrolyzer DN 350 and BDN 250. It is a 1 mm sheet material stretched and half rolled to a thickness of 1.5 mm.

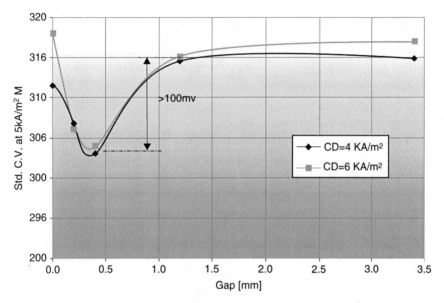

Figure 5.11 Cell voltage as a function of electrode gap [83]. This figure shows the correlation of cell voltage with electrode gap using DuPont N982 membranes, standard Louver and expanded metal electrodes in a 0.16 m² cell—the size of one compartment of the industrial BM2.7 cell.

such as Raney nickel, others incorporate a platinum group metal [4]. The overpotential of activated cathodes is 50–100 mV.

Electrode spacing

As expected, the space between the electrodes and the membrane is a significant factor in determining cell voltage (Fig. 5.11). It is common practice to gently force the membrane against the anode by a slightly higher pressure on the catholyte side of the membrane.

5.1.2.5 Process Description

Because of the extreme sensitivity of high-performance chlor-alkali membrane to brine impurities, brine purification is an important step prior to the electrolysis of sodium chloride by the membrane process. Calcium, magnesium and sulfate ions are almost always present in objectionable quantities. The maximum allowable limits for calcium plus magnesium are 0.025 ppm, and for sulfate ions about 5000 ppm.

The mechanism of membrane damage is in all cases the precipitation of insoluble compounds. In an operating chlor-alkali membrane there is a gradient of OH ion concentration, actually extending into the boundary layer of stagnant electrolyte on both sides of the membrane: with the bulk anolyte at a pH of about 3, the boundary layer of anolyte is close to neutral. This is where the most insoluble hydroxides, such as ferric hydroxide, precipitate. Inside the membrane the concentration of OH ions increases from the anolyte side to the catholyte side, gradually in the sulfonate layer and more steeply in the carboxylate layer. The slope of the gradient is determined primarily by the ability of the carboxylate layer to reject OH ions, that is, the CE. The OH ion concentration in the catholyte boundary layer is higher than in the bulk catholyte because a large fraction of the current in the catholyte is carried by OH ions.

According to this gradient, magnesium hydroxide will precipitate in the sulfonate layer close to the anolyte surface and calcium in the carboxylate layer.

The brine purification process begins with the removal of sulfate ions by precipitation with barium chloride, or less expensively with calcium chloride. The required calcium chloride can be obtained by dissolving the calcium carbonate precipitated in the subsequent step in hydrochloric acid. More recently, removal of sulfate ions by nanofiltration has been suggested [5]: nanofiltration is a process related to reverse osmosis, except that a slightly more open pore structure allows the passage of monovalent ions but rejects ions, such as sulfate ions, carrying a multiple charge. As a result, the nanofiltration of a concentrated raw brine requires much less pressure than customary in reverse osmosis and yields a purified brine with a acceptable low level of sulfate ions.

The bulk of the calcium and magnesium are precipitated as hydroxides and carbonates by the addition of sodium hydroxide and sodium carbonate. The few ppm of calcium and magnesium remaining are removed by an ion exchange process. The resins utilize dibasic acid exchange site to achieve a very high selectivity for divalent ions even in concentrated solution of monovalent salts. Suitable functional groups are amino methyl phosphonic acid and imino diacetic acid groups. The resins are available from Bayer as Lewatit OC 1060 and Lewatit TP 208, respectively, and from Rohm & Haas as Duolite C-467 (also available from Sigma). Strontium and barium are also absorbed by these resins but less strongly than calcium. They will, therefore, break through the bed first and should be the indication for regeneration. Regeneration of these resins is effected by acidification, which converts the functional groups to the non-ionized acid, washing and neutralization with NaOH. The neutralization converts the functional groups back to the hydrated ionic form, resulting in about 35% volume expansion of the bed. Because the depleted anolyte is saturated with chlorine gas and will contain some sodium hypochlorite and chlorate, removal of these oxidizing agents is needed for the protection of the ion exchange resins. A small side stream of the depleted anolyte is strongly acidified with HCl and kept hot in a hold-up tank to effect chlorate destruction (this reaction is not as fast as the hypochlorite destruction). After the chlorate destruction this stream is recombined with the main stream to use the residual acidity for hypochlorite decomposition. The bulk of the chlorine generated in these steps together with the originally dissolved chlorine is removed by vacuum and remaining traces are reduced with sodium sulfite or bisulfite.

While calcium and magnesium are the most common brine contaminants and are damaging even at very low concentrations (0.025 ppm), strontium and barium are encountered less frequently and can be tolerated at higher levels (0.4–0.6 ppm) because of the higher solubility of their hydroxides. Other brine contaminants that are encountered in some locations include iodine, silicic acid and aluminum. Iodides and other iodine compounds may be a contaminant in the salt. These compounds are oxidized to iodates in the anode compartment of the cell. At the higher pH encountered inside the membrane, they can be further oxidized to periodates, possibly by hypochlorites present. Sodium or calcium periodate may then precipitate inside the membrane. Maximum iodine concentration in the brine is 1 ppm at CDs below 4 kA/m^2 and 0.2 ppm at higher CD. Silicic acid and aluminum are converted to anions inside the membrane as the pH increases. The electric field then drives them back in the direction of the anolyte, resulting in "focussing" at the pH of their isoelectric point. Maximum impurity levels are 10 ppm (as silica) for silicic acid and 0.1 ppm for aluminum.

The control of the anolyte chemistry is important for membrane life and the overall efficiency of the process. The anolyte pH is the key factor in this respect. It must not drop below 2 (a value of 2.5 is safer) otherwise irreversible membrane damage is to be expected due to the protonation of the carboxylic barrier layer. Even the most efficient membranes will allow a few percent of hydroxide backmigration, which tends to raise the anolyte pH. On the other hand, even the best anode catalyst will allow a fraction of a percent oxygen evolution (instead of chlorine), which releases protons into the anolyte, lowering the pH. At the same time, the hypochlorite equilibrium tends to buffer the pH around 3–4: $Cl_2 + H_2O \leftrightarrow HClO + H^+ + Cl^-$. Hypochlorous acid can further convert to chlorate ion by anodic oxidation or by disproportionation: $3HClO \rightarrow HClO_3 + 2HCl$, generating additional acidity. It is general practice, to add hydrochloric acid to the purified, saturated brine feed at such a rate, that an anolyte pH of slightly higher than 2.5 is maintained. This will neutralize hydroxide ions migrating through the membrane and minimize hypochlorite formation.

At this pH, current transport by protons is not a significant factor and the pH inside the membrane is maintained at a high value due to the back flow of hydroxide ions. This region of high pH actually extends into the stagnant boundary layer of anolyte at the membrane interphase. As the cations of the anolyte are driven by the current through the membrane, they experience a profile of increasing pH. Those cations which form hydroxides of limited solubility will, therefore. precipitate according to the solubility of their hydroxides: ferric hydroxide will precipitate in the stagnant boundary layer and form a rust-colored deposit on the surface of the membrane, which is more or less harmless. Magnesium hydroxide will precipitate inside the sulfonic layer of the membrane and cause an increase in cell voltage without much effect on CE. The more soluble calcium hydroxide will not precipitate until it has reached the carboxylate layer. Here it experiences not only a steep increase in pH, but also a lower water content of the less hydrated polymer, particularly under the dehydrating effect of the concentrated catholyte. Calcium, even in concentrations of a fraction of a ppm, will then precipitate as the hydroxide inside the carboxylate layer or at the carboxylate/sulfonate boundary. Such deposits of calcium hydroxide will not only increase cell voltage, but also decrease CE due to the disruption of the barrier layer. As a result, calcium is one of the most destructive brine contaminants. Trace amounts of barium on the other hand can pass through the membrane into the catholyte harmlessly without precipitating.

Electrolysis conditions

CD:	2–6 kA/m^2	Can be cycled to take advantage of off-peak rates.
		Higher CD will result in less chloride in NaOH.
Temperature:	80–95°C	
Anolyte:	170–230 g/l NaCl	Water transport decreases from 4.6 to 3.1 mol/Na$^+$ within this range
Catholyte:	30–35% NaOH	32 % is optimum. Voltage increases 25 mV for each % increase in NaOH concentration

Membrane damage

Membrane damage is frequently due to brine impurities. Other causes include upset in operating conditions, particularly catholyte concentration above 36% NaOH, anolyte concentration below 170 g/l NaCl, anolyte pH below 2 and mechanical damage (Fig. 5.12). The damage in this picture is believed to be due to abrasion by the cathode, caused by wide variations in the pressure differential between the two compartments.

As the cations move through the membrane, they drag along a certain amount of water of hydration. This is desirable because removal of water from the anolyte reduces the amount of depleted anolyte that needs to be resaturated. A transport of about 5.5 mol of water per sodium ion would be optimum because it is the amount necessary to form 32–34 % sodium hydroxide in the cathode compartment without any additional water feed to that compartment. Actual water transport is about 4 mol/Na$^+$.

The operating temperature of the cells is maintained near 90°. In the range of 75–90°C, the cell voltage drops about 8 mV for each degree of temperature increase. Any further increase in operating temperature would result in a prohibitive increase in the volume of the gas release, as the boiling point of the electrolyte is approached.

Development of the BM 2.7 electrolyzer is continuing. Version 5, planned for introduction in 2007, is designed for a power consumption of less than 2000 kWh/t of NaOH at a CD of 6 kA/m^2. The NaCl content of the 32% NaOH will be less than 10 ppm.

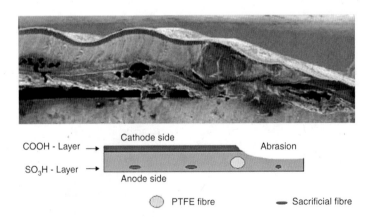

Figure 5.12 Cross-section through a damaged chlor-alkali membrane [82].

BM2.7v3	BM2.7v4	BM2.7v5
1997	2004	2007
• PC ~ 2,200 kWh/t NaOH at 6 kA/m^2, N982	• PC <2,100 kWh/t NaOH at 6 kA/m^2, F8020	• PC <2,000 kWh/t NaOH at 6 kA/m^2, F8020
• NaCl in Caustic <20 ppm in 32% NaOH	• Flooded element operation	• Modified cell frame system
• Laser welded, ribs, downcorner 17x spacer rows	• Minimal internal dp fluctuation	• New spacer system
	• NaCl in Caustic <15 ppm in 32% NaOH	• Material savings
• Metallic distribution pipes	• PTFE distribution pipes	• NaCl in Caustic <10 ppm in 32% NaOH
• Differential pressure at 40 mbar	• Differential pressure at 20 mbar	
• Narrow gap of 1.2 mm	• Very narrow gap of 0.4 mm	

Figure 5.13 Performance of the BM 2.7 electrolyzer ver. 3, 4 and 5 [83].

The performance improvement of the BM 2.7 electrolyzer versions 3, 4 and 5 is shown in Fig. 5.13.

Gas-diffusion cathodes or electrodes (GDCs or GDEs)

High-purity hydrogen is a by-product of the chlor-alkali electrolysis. In some locations, its inherent value can be realized by using it for hydrogenations and other chemical reactions. In other locations, it can only be used as boiler fuel. Efforts have been made to use this hydrogen in fuel cells to supply part of the DC power needed in the electrolysis.

A different approach would be to use only the GDC of a fuel cell as a replacement for the cathode of a conventional chlor-alkali cell. While the fuel cell approach would require an *additional* pair of electrodes, the use of a fuel cathode would *replace* the conventional hydrogen evolving cathode by an oxygen consuming GDC. This will not only minimize the cost of the electrodes, but also eliminate the need for other fuel cell hardware, current collectors and conductors. The fuel cell approach would also suffer the penalty of two additional overvoltages, while the GDC approach will replace the overvoltage of the conventional cathode with that of a GDC. A voltage saving of about 1 V could, therefore, be realized compared to about 0.7 V for a fuel cell. This voltage saving would have to be balanced

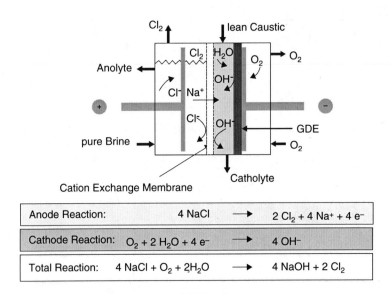

Figure 5.14 NaCl electrolysis with a GDE [82].

by the value of the lost hydrogen production and the cost of the oxygen consumed.

The principle of operation is shown in Fig. 5.14 (GDE). The anode side and the ionomer membrane are identical with that of a conventional chlor-alkali cell. The GDE separates the catholyte in front from the gas phase in the back. The GDE is a three-dimensional electrode that allows, through the use of a catalyst on a high surface area support, to pack a large surface area of electrode into a much smaller geometrical area. A schematic cross-section of a GDE is shown in Fig. 5.15. A catalyst particle is effective only if it is in contact with an electronic conductor reaching back to the current distributor and a continuous electrolyte path leading forward to the bulk electrolyte. It must also have access to oxygen. The GDE, therefore, consists of an intermeshing network of three continuous phases: an electronic conductor, usually formed by carbon black, the liquid electrolyte and a passage for gas. The gas passage is usually provided by rendering certain areas inside the GDE non-wettable. A delicate balance between hydrophilic and hydrophobic areas is required to establish the gas–electrolyte meniscus at the optimum location. A wire screen made of nickel, or silver plated nickel, is used as the current distributor and also provides mechanical support for the GDE. pTFE fine powder is used as a binder for the supported catalyst and also provides hydrophobic gas passages. On mechanical working or shearing composition containing pTFE fine powder,

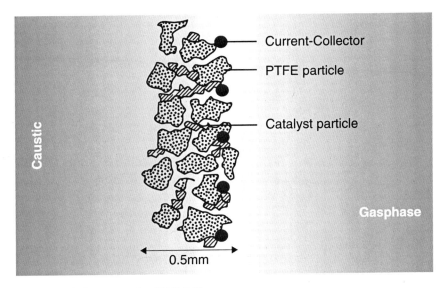

Current-Collector

PTFE particle

Catalyst particle

Caustic

Gasphase

0.5mm

Figure 5.15 Structure of a GDE [82].

the pTFE particles tend to fibrillate, thereby providing long, continuous hydrophobic gas passages.

Maintaining the electrolyte/gas meniscus at the proper location within the GDE is made more difficult by the hydrostatic pressure at the bottom of the cell. One approach to maintain a balance of gas and electrolyte pressure has been to subdivide the gas compartment in the vertical dimension into gas "pockets" maintained at different pressures. Another approach has been to replace the free-standing catholyte with a catholyte imbibed within a "percolator" that limits the downward flow of catholyte without generating a hydrostatic pressure on the bottom.

A study of the degradation of GDC used in chlor-alkali membrane cells has been published recently: Okajima et al., J. El. chem. Soc. **152**(8), pp. D117–D120, Aug. 2005.

5.1.3 Related Technology: HCl electrolysis

The use of chlorine in organic synthesis frequently results in the generation of by-product HCl. The conversion of such by-product HCl to chlorine by electrolysis requires less voltage than the electrolysis of salt: the decomposition voltage is lower because the hydrogen is evolved from a strongly acidic medium instead of 32% NaOH, and the cell resistance is lower because of the high conductivity of aqueous HCl. The anolyte and the catholyte are both approximately 21% aqueous HCl, a concentration that combines maximum conductivity with minimum vapor pressure. A separator serves to separate the gases (chlorine and hydrogen) evolved at the two electrodes.

Before the advent of membrane technology, PVC fabric was used in combination with bipolar graphite electrodes for the commercial process. The gas separation was not perfect, in part because the electrolytes percolating through the fabric contained dissolved gases. The use of a perfluorinated ionomer membrane provided more complete gas separation, and therefore gas purity, and higher conductivity and longer life than the PVC fabric. A further reduction in cell voltage is possible through the use of a GDE or oxygen depolarized cathode (ODC) as illustrated in Figs 5.16–5.18.

5.1.4 The Older Technologies: Mercury and Asbestos

For comparison purposes, a brief review of the two older technologies, not using ion exchange membranes, may be useful.

In the mercury or amalgam cell, a cathode of flowing mercury is used. Sodium is deposited in this cathode in the form of sodium amalgam. Ordinarily, the cathodic deposition of metallic sodium is not possible in an aqueous system, because the standard potential of sodium is very negative relative to hydrogen and hydrogen evolution is the only cathode reaction.

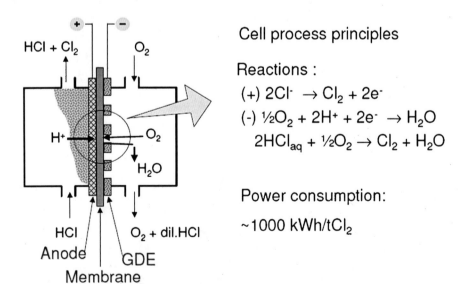

Cell process principles

Reactions :

(+) $2Cl^- \rightarrow Cl_2 + 2e^-$

(-) $\frac{1}{2}O_2 + 2H^+ + 2e^- \rightarrow H_2O$

$2HCl_{aq} + \frac{1}{2}O_2 \rightarrow Cl_2 + H_2O$

Power consumption:

~1000 kWh/tCl$_2$

Figure 5.16 HCl electrolysis using a GDE [83].

Membrane bipolar cell

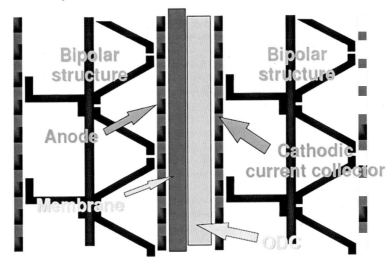

Figure 5.17 Cell construction using an ODC [83].

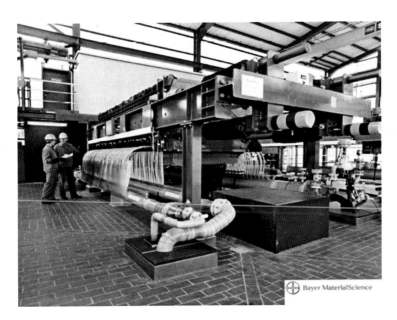

Figure 5.18 HCl electrolysis using an ODC at brunsbuettel; 50 t/day capacity [83].

The formation of sodium amalgam is made possible by two factors. the potential necessary to form sodium amalgam is about 0.8 V less negative than that required to form sodium metal (this energy difference between sodium metal and sodium amalgam is responsible for the spectacular fireworks displayed when sodium amalgam is made by combining the two elements). The other factor is the very high overvoltage for the discharge of hydrogen on a mercury cathode. This overvoltage stabilizes sodium amalgam in the presence of water, by allowing the amalgam to acquire its negative equilibrium potential.

The sodium amalgam, dissolved in excess mercury, is decomposed in an external "denuder", consisting of a bed of graphite particles immersed in a 50% solution of sodium hydroxide. As the liquid amalgam flows over and makes electrical contact with the graphite particles, the equilibrium potential which the amalgam acquires by sending a few sodium ions in solution, is sufficiently negative to cause discharge of hydrogen on the low overvoltage graphite. The denuder, therefore, acts as a short circuited fuel cell: a negative amalgam electrode (anode) generating electrons by releasing sodium ions into solution and a graphite cathode using the electrons it receives from the mercury to generate hydrogen gas.

The main advantage of the mercury cell is that the sodium hydroxide generated is of both high purity and concentration. The disadvantages are: (1) A high cell voltage because of the more negative potential required for amalgam formation relative to the discharge of hydrogen. This extra electrical energy is released in the denuder as heat. (2) Large requirement for floor space because of the horizontal electrodes. (3) Environmental problems because of the use of mercury. Historically, mercury technology has dominated the chlor-alkali industry in Europe and Japan. In Japan, it has been completely eliminated and replaced by membrane technology. In Europe, a considerable effort has been made to solve the environmental problem associated with this technology and mercury emission have now been reduced to less than 2 g of mercury per ton of chlorine produced. Emissions in the products (lower bar) and water (middle bar) have been almost completely eliminated (Fig. 5.19).

Nevertheless, the share of mercury technology in European chlor-alkali production has substantially decreased during the past 30 years (Fig. 5.20).

The asbestos cell uses an asbestos diaphragm deposited directly on the wire screen cathode by a paper making technique. Asbestos does not have inherent conductivity; the diaphragm gains conductivity only through the imbibed electrolyte. This diaphragm is, therefore, unable to prevent the backflow of the highly mobile OH ions (compare Table 5.1). This backflow is overcome by pumping anolyte through the diaphragm at a rate

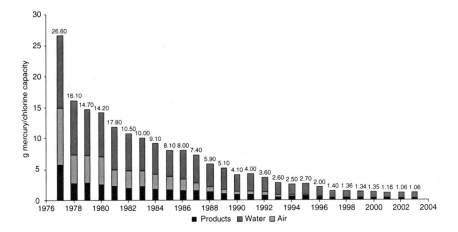

Figure 5.19 Western European mercury emissions [84]. Note: East European members are not part of the voluntary commitment yet but are shown for completeness.

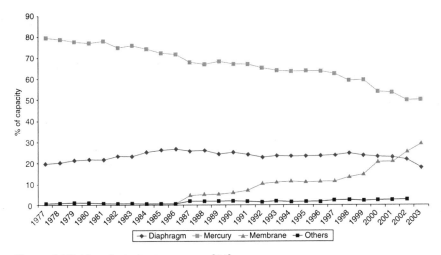

Figure 5.20 Manufacturing processes [84].

sufficient to balance the OH ion migration. As a result, a catholyte product containing in addition to 11% sodium hydroxide, 16% unconverted sodium chloride is obtained. The expense of concentrating and purifying this raw product is the main disadvantage of the asbestos technology. The asbestos technology is still the dominant technology in North America and is only slowly being replaced by membrane technology.

5.1.5 Electrolysis of Other Chlorides, Including Tetramethyl Ammonium Chloride

The electrolysis of potassium chloride is easier than that of sodium chloride because the less hydrated potassium ion results in a less swollen, and therefore more selective, membrane. In addition, the mobility of the potassium ion is higher than that of the sodium ion (see Table 5.1). As a result, high CEs are obtained even with membranes without any barrier layer. However, the less hydrated membrane has a higher electrical resistance, making inherent membrane conductivity an important factor. Nafion 551, with its very open reinforcement, is therefore the preferred membrane for this application.

Because of the lower hydration of the membrane in the potassium form, differences in the operating conditions compared to the NaCl electrolysis are to be expected. The anolyte concentration should be lower than in NaCl electrolysis: the optimum is 130–140 g/l KCl with an allowable range of 130–200 g/l. For a given anolyte concentration, expressed in g/l, the water transport in KCl electrolysis is lower than in NaCl electrolysis. The difference is even more pronounced if the concentration is expressed in mol/l. However, at the optimum anolyte concentration of 135 g/l KCl vs. 200 g/l NaCl, the water transport is almost the same (Fig. 5.21).

The CE is the highest at a catholyte concentration of about 34% KOH for the all-sulfonate membranes, and 30% KOH for membranes with a

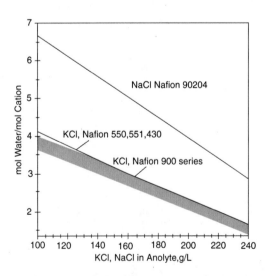

Figure 5.21 Comparison of water transport in KCl vs. NaCl electrolysis [85]. Conditions: 3.1 kA/m², 90°C, 32% KOH or 32% NaOH.

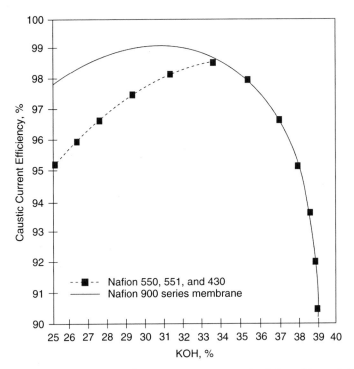

Figure 5.22 Current efficiency vs. KOH concentration [85]. Conditions: 3.1 kA/m²,
90°C, and 110 to 200 g/L exit brine.

carboxylic barrier layer (Fig. 5.22). Figure 5.23 shows the cell voltage
as a function of CD for various types of Nafion used in the electrolysis
of KCl.

The electrolysis of lithium chloride [6], on the other hand, is more
difficult. The equivalent conductivity of the lithium ion is much smaller
than that of either sodium or potassium. Furthermore, the larger hydration
shell of the lithium ion results in a more swollen, and therefore less selective, membrane. This factor is aggravated by the low solubility of lithium
hydroxide, even hot, which limits the dehydrating effect of a concentrated
catholyte. Back migration of hydroxide ions, therefore, limits the CE to
about 70%.

The manufacture of tetra alkyl ammonium hydroxides, particularly tetra
methyl ammonium hydroxide, by electrolysis of a solution of their salts
has received considerable attention and is practiced on a commercial scale.
Wade et al. report on the electrolysis of tetra methyl ammonium chloride
[7]; see also Ref. [8]. To prevent chlorination of the alkyl ammonium ion,

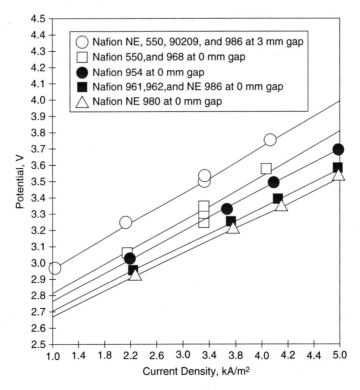

Figure 5.23 Cell voltage vs. CD for KCl electrolysis [85]. Conditions: 90°C, 32% KOH, 170 to 200 g/L KCl (act. NI cathodes for 0 mm gap only).

chlorine scavengers, such as methanol, may be added to the anolyte. Or the bicarbonate or formate of the tetra alkyl ammonium ion is used instead of the chloride [9,10].

5.1.6 Potassium Gold Cyanide and Potassium Stannate

These are two examples of the electrochemical production of electroplating chemicals. Potassium gold cyanide is the most important gold plating chemical. Almost all the gold used for contacts, etc. in the electronic industry is derived from potassium gold cyanide. The older route to $KAu(CN)_2$ started with the dissolution of metallic gold in aqua regia. It was followed by a lengthy procedure to replace unwanted chloride ions with cyanide ions and to remove excess nitric acid. Some losses of gold compounds were unavoidable and high product purity could not always be assured.

Anodic dissolution of gold eliminates the introduction of any unneeded chemicals.

The anode reaction is: $Au + 2KCN \rightarrow KAu(CN)_2 + K^+ + e^-$
The cathode reaction is: $K^+ + e^- + H_2O \rightarrow KOH + \frac{1}{2}H_2$

For the anodic dissolution of tin in potassium hydroxide: $Sn + 6KOH \rightarrow K_2SnO_3 + 4K^+ + 4e^- + 3H_2O$. The cathode reaction is again the formation of potassium hydroxide and hydrogen. A divided cell is required in both cases to prevent the product formed in the anode compartment from plating out at the cathode.

The commercial process uses an open tank cell. The tank is filled with anolyte into which anode baskets are suspended from overhead bus bars. The baskets are filled with granular gold or tin ingots, respectively, and fresh metal is added to the baskets as needed. Also, suspended in the anolyte are cathode bags made of Nafion and filled with catholyte. Stainless steel or nickel can be used as cathode material. Any contact of the membrane with the cathode must be avoided.

Both reactions generate by-product potassium hydroxide in the cathode compartment. To maintain a reasonable concentration of KOH in the catholyte, it is necessary to remove part of the catholyte and replace it with water either continuously or intermittently.

Potassium gold cyanide is usually made in a batch process. If the amount of current passed through the cell is approximately equivalent to the initial charge of potassium cyanide in the anolyte, the anolyte at the end of the reaction contains very little contaminants and will yield a very high-purity product on cooling. The mother liquor is reused for the next batch. Losses of gold are thereby minimized.

Nafion 424 is the membrane of choice for both reactions. In the case of potassium gold cyanide, the membrane is used in "reverse orientation". This means that although the cathode is inside the bag, the bag is made with the cathode surface of the membrane outside. The reason for this is that brown contaminants in the reinforcing fabric can leach out of the anode surface of the membrane and the reverse orientation, therefore, avoids contamination of the product in the anolyte.

Trace amounts of gold compounds will migrate through the membrane, particularly at low or zero CD or through areas of the membrane that for geometrical reason receive little current. These compounds will plate out on the cathode, usually in the form of a fine powder, which falls off and accumulates in the bottom of the bag. There it may act as a parasitic bipolar electrode: the part touching the membrane acts as a cathode and the back, facing the cathode, will act as an anode. Under these circumstances, dendrites growing from the cathode side of the powder can puncture the membrane and then grow rapidly in the anolyte. Compare with a similar problem discussed at the end of Section 5.1.8. The insertion of a plastic strip folded to a "V" into the bottom of the bag can overcome this problem.

5.1.7 Salt Splitting

Electrochemistry allows the splitting of salts into their acid and base components. In the case of using a cation exchange membrane for this process, the cations (such as Na^+ or K^+) of the salt dissolved in the anolyte are forced by the electric current through the membrane into the catholyte. The cations in the anolyte are replaced by hydrogen ions generated by the anodic oxidation of water. The electrode products (hydrogen and oxygen) are inadvertent by-products of this process. The main advantage of the electrochemical process compared to conventional acidification using sulfuric acid is the elimination of by-product sodium sulfate. As a result, the acid is obtained in higher purity, and yield losses involved in the separation of sodium sulfate from the product acid are avoided. The process is applicable to any acid, particularly weak acids, which are manufactured from their salts. Examples are chromic, boric, and salicylic acids. See also Section 5.3.

This process is used on a commercial scale in the production of chromic acid from sodium dichromate. It is interesting that the main driving force for the adoption of the new technology was again environmental: the disposal of sodium sulfate contaminated with even traces of chromic acid was extremely difficult. One could say that the poor purity and yield of the main product was not as serious a problem as the poor purity of the waste product!

The process of making chromic acid begins with the alkaline roasting of chrome ore in a rotary kiln. Air is used to oxidize the trivalent chromium in the ore to the hexavalent state in the form of sodium chromate. This is leached with water to yield a concentrated solution of sodium chromate. Half of its sodium content can be removed inexpensively by acidification with carbon dioxide under pressure. While carbon dioxide is a rather weak acid, the process is driven by the low solubility of the sodium bicarbonate formed. The sodium bicarbonate removed by filtration and is converted to sodium carbonate by heating. The sodium carbonate is fed back into the rotary kiln, closing one half of the "alkalinity loop".

At this point the electrochemical route departs from the older technology (acidification using sulfuric acid): the sodium dichromate solution is fed into the anode compartment of an electrolytic cell. The electric current forces sodium ions through the membrane into the cathode compartment where they could form sodium hydroxide just like in the electrolysis of NaCl. However, the low pH of the anolyte precludes the use of carboxylic/sulfonic bimembranes, which allow the production of 30–35% NaOH at high CE. Sulfonic membranes in comparison produce 15% NaOH at about 85% CE. To avoid the expensive evaporation necessary to feed such a dilute solution back into the "alkalinity loop", it is preferred to use a solution of sodium dichromate as the catholyte (i.e., sodium dichromate solutions are used as feeds for both anolyte and catholyte) [11]. As sodium ions move through the membrane into the cathode compartment, the catholyte is gradually converted to sodium chromate and the anolyte to chromic acid. Sodium chromate formed in the catholyte is then recycled back to the carbon dioxide acidification, which yields alkalinity in solid form. This closes the second half of the "alkalinity loop".

The CE of this process is very high up to the point, where the sodium content of the anolyte is depleted so much that hydrogen ions participate in the transport of current. The anolyte flow is, therefore, cascaded from one cell to the next, so that the penalty of low CE is suffered only in the final stages of the cascade. When the desired degree of sodium ion removal has been achieved, the anolyte is discharged from the cell and subjected to partial evaporation. On cooling, high-purity chromic acid crystallizes. The mother liquor, which contains chromic acid and the residual sodium ions in the form of sodium dichromate or tetra chromate, is fed back into the electrolytic cascade at a point where the chromium/sodium ratios match.

If the membrane touches the cathode, membrane damage may occur, eventually resulting in chromium metal plating on the membrane surface in the anode compartment. Compare with the discussion at the end of Section 5.1.8.

Sodium sulfate is produced on a very large scale as a by-product of several important industrial processes. In many cases, disposal of this material is difficult. As a result, there have been many efforts to use electrolytic salt splitting to recover both sulfuric acid and sodium hydroxide. Because of major technical and economic problems, none of these efforts have been successful. The technical problems are caused by the acid strength of sulfuric acid:

1) Even a small conversion of sodium sulfate to sodium bisulfate lowers the pH of the anolyte below 2, which precludes the use of high-efficiency membranes containing a carboxylic barrier layer. The use of a sulfonic barrier layer

(Nafion 324) limits the maximum caustic concentration generated as the catholyte to about 15%.

2) Competition by hydrogen transport: as sodium ions in the anolyte are removed by current transport through the membrane, they are being replaced by hydrogen ions generated by the anodic oxydation of water with the release of oxygen gas. At first, most of these hydrogen ions are captured by sulfate ions to form bisulfate ions, which are not completely dissociated in the presence of excess sulfate ions. At these low conversions, loss of CE due to transport of hydrogen ions through the membrane is therefore not a serious problem. As the anolyte is depleted of sulfate ions, the dissociation of bisulfate becomes more pronounced and at the point of complete conversion of the anolyte to sodium bisulfate, that is, when one half of the original sodium content of the anolyte has been removed, CE drops to an unacceptable low value. The use of a three-compartment cell with the sodium sulfate feed first entering the center compartment and then overflowing to the anode compartment can extend the maximum practical conversion to about 60%. Alternatively, with a sufficiently high flow through the center compartment, it may be possible to maintain its pH above 2.5, thereby allowing the use of a high-efficiency membrane between the center and the cathode compartments.

The economic problems are related to the cost comparison with sodium hydroxide generated by the electrolysis of salt (NaCl): while the operating costs of the two processes are almost the same, the NaCl electrolysis generates chlorine as a valuable by-product. The value of oxygen generated by salt splitting is comparatively minor.

In view of these considerations, the following conditions must be met in order for the electrolytic splitting of sodium sulfate to be successful:

1) Both acid (in the form of a sodium bisulfate solution) and caustic (in the form of a 15% NaOH solution) must be reusable on site with full or almost full credit for their value.

2) Sodium sulfate must be available at a negative value; that is, there must be some credit for disposing an undesirable by-product.

5.1.8 Chromic Acid Regeneration

In many industrial processes using chromic acid, a spent acid stream containing trivalent chromium and/or cationic contaminants is obtained.

Disposal of this stream would require chemical reduction of any remaining hexavalent chromium followed by precipitation of chromium hydroxide and filtration. Electrochemical regeneration is an attractive alternative. It is being used on spent acid streams generated in three applications:

1) Chromic acid used in the oxidation of some organic compound.
2) Chromic acid used in surface etching of plastic parts prior to metallizing.
3) Chromic acid plating solutions.

In all the cases, anodic oxidation of trivalent to hexavalent chromium is the main desired reaction. To achieve high anodic CE, selection of the anode material is most important. Lead dioxide is the material of choice. Low anodic CD and high anolyte turbulence also contribute to high CE, even as the concentration of trivalent chromium is reduced to low levels. At the membrane, removal of cationic contaminants is desired; however, because their concentration is typically low, poor membrane CE is to be expected with most of the current carried by hydrogen ions. If these cationic contaminants are sufficiently noble, such as copper, they may plate out on the cathode, but again at poor CE. Contact of the membrane with the cathode should, therefore, be avoided. The main cathodic reaction is the evolution of hydrogen.

The catholyte for the first two cases is dilute sulfuric acid. In the last case, the choice of catholyte is more limited, because in chrome plating, the content of sulfate ions in the plating bath must be controlled within narrow limits. Because the anion rejection of the membrane is not absolute, sulfuric acid of the typical concentration range (10–20%) may result in excessive sulfate contamination of the plating bath. To avoid this problem, anions such as carbonate, formate or citrate have been used in the catholyte.

For this application, a cylindrical cell design offers several advantages: (1) The cell is installed in a corrosion proof container without any external gaskets; the possibility of a chromic acid spill is, therefore, greatly reduced. (2) The external anode has more area than the membrane and the cathode. This results in low anodic CD, which promotes high anode efficiency. (3) The centrally located cathode can be pulled easily for cleaning. A diagram of such a cylindrical cell is shown in Fig. 5.25, and the components of the cathode compartment in Fig. 5.24: on the left is the entire cathode compartment consisting of an outer titanium cage supporting the conical bottom closure made of Teflon. The three components of the top closure, designed very similar to the bottom closure, are shown in the center of the bottom of the picture.

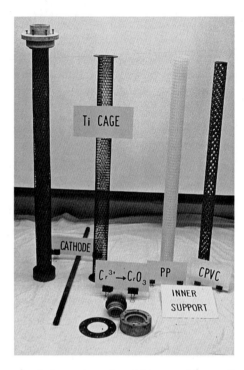

Figure 5.24 Components for a chromic acid regeneration cell.

An outer sleeve made of Teflon has a conical inner opening. The conical Teflon plug, with its three VITON O-rings, has a diameter and taper that matches the conical opening of the outer sleeve. For assembly, the top of the cylindrical membrane is inserted from the bottom into the opening of the outer sleeve. The plug is then inserted into the inside of the tubular membrane and forced into the taper of the outer sleeve by bolting the titanium ring to the top of the sleeve. The dimensions of the tubular membrane are 7.6 cm diameter × 1.2 m long. Two examples of an inner support to prevent membrane–cathode contact are shown on the right of the picture. A schematic and a picture showing an installation of three of these cells in connection with a shared catholyte tank is shown in Appendices A and B.

Ahmed has developed a mathematical model to estimate contaminant fluxes due to migration, diffusion and convection in the electrolytic regeneration of contaminated hard chrome plating baths using Nafion 117 as the separator. The ionic mobilities of Cu, Fe and Ni through Nafion 117 were found to be 5.4×10^{-10}, 1.7×10^{-10} and 5.2×10^{-10} cm^2/Vs, respectively [12].

Trace contaminants such as copper may plate out on the cathode. If the membrane touches the cathode under these circumstances, dendrites

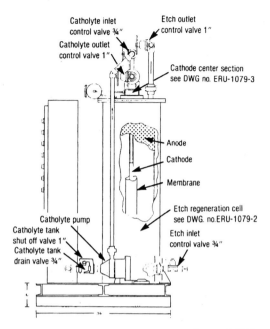

Figure 5.25 Schematic of a chromic acid regeneration cell [86].

growing on the cathode may grow through the membrane. When the tip of the dendrite comes close to the anolyte, rapid plating of chromium metal may occur. It appears that the chromium plates out in nodules on the anolyte surface of the membrane. As in many other electrolytic applications, contact of the membrane with the cathode should, therefore, be avoided.

Because hexavalent chrome plating will not resume even after a very brief (1 s) interruption of current, a periodic interruption of current may minimize this problem.

A chromic acid regeneration system incorporating three of these cells is shown in Appendices A and B.

A laboratory scale cell for chromic acid regeneration is shown in Fig. 9.11.

5.1.9   Nickel–Zinc Plating

In a recent development, a perfluorinated ionomer membrane is used to prevent anodic oxidation of an aliphatic imine additive to the plating bath [13]. The nickel–zinc coating provides superior corrosion protection compared to zinc plating.

5.1.10   Persulfuric Acid and Persulfates

Persulfuric acid and persulfates have been made for many years using a porous separator to prevent cathodic reduction of the peroxy compounds formed. More recently, perfluorinated ionomer membranes have been used as a separator to more effectively separate the two electrolytes and to lower the cell voltage. High conductivity is important in this application, because a high anodic CD is required for good CE and the desired low operating temperature in general will require refrigeration of at least the anolyte. A high cell voltage will, therefore, not only increase the electric power consumption, but also the refrigeration load. Dong [14] suggests the simultaneous production of persulfates in the anode compartment with production of hydrogen peroxide by the cathodic reduction of oxygen in the cathode compartment.

5.1.11   Electro-winning of Metals

Perfluorinated ionomers have been used as a separator in the electro-winning of metals, particularly gold. The purpose of the separator is to prevent interference of oxidants, such as ferric ion, with the cathodic deposition of the metal [15]. While the chemical environment in the electro-winning of metals is usually not very corrosive and may allow the use of non-fluorinated separators, the ready availability of Nafion combined with the ability to fabricate this material into bags and other shapes, has made it the separator of choice in this application.

The Anglo American Research Lab has designed a particularly effective cell for the cathodic deposition of gold from a cyanide leach solution, even at concentrations as low as 0.1 ppm. In this case, an additional purpose of the separator is to prevent the corrosion of the 316 stainless steel anode by the chloride contaminant in the catholyte and to prevent anodic oxidation of the cyanide values in the leach solution. Also, the separator prevents oxygen generated at the anode from entering the cathode compartment, where it would interfere with the deposition of gold.

The cell is cylindrical with the anode inside surrounded by a tube made of Nafion. The dimensions of this tube are 32 cm diameter × 100 cm long (or high). The outside of the Nafion is protected by a plastic mesh onto which a packing of about 1.7 kg of steel wool is wound. This in turn is surrounded by a stainless steel current collector. The leach solution serving as the catholyte is pumped vertically (upflow) through the annular space filled with steel wool at a rate of 50–70 l/min. Because the liquid flow is through a relatively narrow cross-section and along a fairly long path, the linear velocity is very high for effective mass transfer from the bulk of the solution to the surface of the steel wool. A 20% solution of sodium hydroxide is used as the anolyte.

The process is typically run batch-wise starting with a fairly high concentration of gold (1500 ppm) and a high current (200–250 A on about 1 m² of membrane area) at 50–60°C. Under these conditions, 20–50% of the gold in solution is removed in a single pass at high CE. As the solution becomes depleted, the current is reduced to 100–130 A to avoid excessive hydrogen evolution. While the cell can deplete the catholyte to a level of less than 0.1 ppm of gold, the process is typically stopped when the concentration of gold has dropped to a few ppm. At least 16 kg of gold can be recovered in a few days on a 1.7-kg packing of steel wool. The depleted catholyte is reused for leaching gravity concentrates or loaded carbon from the carbon-in-pulp process.

In this type of three-dimensional electrode, the potential distribution has to be considered: because the electrolyte is typically a poor conductor, there is a significant potential drop in the electrolyte filling the voids in the steel wool packing. The steel wool electrode in comparison has a high conductivity and therefore a fairly uniform potential along its thickness. The potential difference between electrolyte and electrode is, therefore, highest at the front face of the packing (near the membrane) and lower near the current collector. If the rate of metal deposition at the front face is mass transfer limited, hydrogen evolution may occur at the front while metal deposition in other parts of the electrode is limited by the lower potential difference. Increasing the current under these conditions will mostly increase the rate of hydrogen production. Increasing the temperature and/or adding a supporting electrolyte, on the other hand, will increase electrolyte conductivity and therefore CE. Or the addition of non-conductors to the steel wool packing can increase the electrical resistance of the packing and match it to the electrolyte resistance.

5.1.12 Water Electrolysis (Including the Production of Ozone or Hydrogen Peroxide)

Water electrolysis cells using ionomer membrane technology avoid the use of dissolved electrolytes and the associated corrosion and contamination problems. Pure water is the only feed to the electrolyzer. These are frequently designed similar to PEM fuel cells. The membrane is contacted directly by gas permeable electrodes on both sides. There are three methods available to provide the electrode/ionomer contact:

1) A catalyst can be precipitated on the surface or inside a thin surface layer of the ionomer film by chemical reduction [16,17]. For this the ionomer is converted to an ionic form of the catalytic metal, and then immersed in a solution of a reducing agent that is unable to diffuse into the ionomer. Sodium borohydride is suitable, because the reducing component is an anion. If a cationic reducing agent, such as a hydrazonium salt or stannous chloride, or some neutral compound such as formaldehyde, would be used, metal deposition would occur deep inside the ionomer film. The ion exchange/reduction cycle can be repeated several times for higher loadings.

2) A catalyst "ink" can be prepared and coated on the surface of the ionomer, similar as it is done in fuel cells.

3) The electrode can be pressed mechanically against the ionomer film.

While the physical arrangement of the electrolyzer is similar to that of a PEM fuel cell, there are important differences in the composition of the electrodes: most importantly, carbon black as a catalyst support will be oxidized during anodic oxygen evolution. The most successful anode catalyst has been iridium dioxide. It is used without any support. Ruthenium dioxide will actually give a lower initial cell voltage; however, its performance declines after several hundred hours of operation. Platinum is the most commonly used cathode catalyst.

Takenaka et al. [16] reported a Tafel slope of 40–60 mV/decade for an iridium anode and, in combination with platinum cathode, a cell voltage of 1.56–1.59 V at 500 mA/cm^2 and 90°. Michas and Millet [17] pressed ruthenium dioxide electrodes, made by pyrolysis of ruthenium chloride on a porous titanium substrate, on a Nafion membrane and compared the performance with a platinum/Nafion composite made by sodium borohydride reduction of a platinum-exchanged Nafion film. Holze and Ahn [18] compared the performance and long-term stability of membrane–electrode composites made by various methods. The chemical reduction route (method 1) worked well only for platinum electrodes; all other electrodes, incorporating both noble metals and their oxides, were fabricated from pTFE-bonded catalyst layers pressed onto the membrane. A regenerative electrolyzer/fuel, cell using pTFE-bonded electrodes, was described by Ioroi et al. [19].

Stucki and coworkers have obtained a patent on water electrolyzers using a fluorinated ionomer electrolyte between an anode catalyst consisting of a mixture of 80% iridium, 20% ruthenium dioxide and a finely divided platinum cathode catalyst [20].

The electrolytic generation of ozone using polymer electrolyte separators has been reviewed by Han et al. [21]. A Swiss company Ozonia offered a commercial ozone generator based on the solid polymer electrolyte concept: a 0.25–mm-thick film of 1200 EW Nafion (Nafion 120) was coated on one side with a porous platinum electrode similar to the electrodes used in fuel cells. The anode is made of lead dioxide coated on a porous titanium substrate and is pressed against the other side of the film. Pure water is fed to the back side of the titanium substrate (Fig. 5.26) [22–24]. Because the ionomer film is the only electrolyte in the system, the cell can

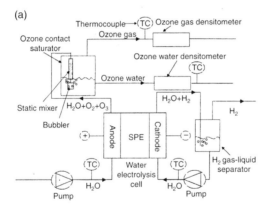

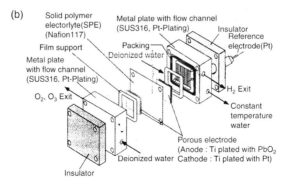

Figure 5.26 Electrolytic generation of ozone. Schematic diagram of (a) experimental apparatus, and (b) water electrolysis cell.

deliver a solution of ozone in water directly. Or the ozone can be obtained as a gas. The cathode can be depolarized with air to minimize the required cell voltage [25]. More recently, Lynntech International has been granted patents in this area [26]. Onda et al. published an extensive study of the various factors affecting the efficiency of ozone production using Nafion 117 between a lead dioxide anode and a platinum cathode, both plated on titanium: electrode construction, flow rate and temperature of the water, CD, and the use of a trickle current to prevent the performance degradation due to current interruption [27]. The results were compared with the earlier work by Stucki and coworkers.

Stucki described an electrolytic cell for the production of hydrogen peroxide solutions by cathodic reduction of oxygen. The cathode consists of activated carbon (charcoal) or certain metal chelates. The preferred anode catalyst is a mixed iridium/ruthenium dioxide. The oxygen generated at the anode can be used as a feed for the cathode. The ionomer film can, therefore, be gas-permeable and serves only as a barrier to the flow of liquids and electrons [28]. Ando and Tanaka reported on the simultaneous production of hydrogen and hydrogen peroxide by the electrolysis of water in a cell equipped with a Nafion membrane [29].

5.1.13   Other Electrolytic Applications

Listed here are applications of technical interest, even though they may not be in commercial use today. They may serve to illustrate the broad range of product that can be made by electrolytic processes.

HFPO has been made by anodic oxidation of HFP in an anolyte typically containing about 84% acetic acid, 10% nitric acid, 4% water and 2% HF. The cylindrical cell consists of an outer shell made of stainless steel, which had been coated on the inside with lead dioxide. This forms the anode. The stainless steel cathode is in the center; a sleeve made of Nafion 427 (now replaced by Nafion 424) separates the anode and the cathode compartments [30,31].

Olin Corporation published several electrolytic processes at the 11th International Forum on Electrolysis in the Chemical Industry, Nov. 2–6, Clearwater Beach Florida:

1) Sodium hydrosulfite, also known as sodium dithionite, can be made by cathodic reduction of sodium bisulfite (ECHO Process, Fig. 5.27). Important for the success of this reaction is the use of a high surface, multilayer stainless steel fiber cathode. This allows a low CD at the cathode for

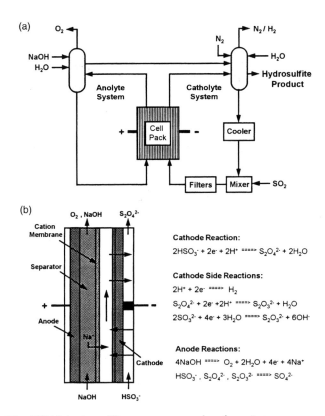

Figure 5.27 ECHO hydrosulfite process overview (*courtesy*: The ElectroSynthesis Co. [87]). (a) Overview of ECHO hydrosulfite process. (b) Chemistry of ECHO process electrolysis.

efficient reduction combined with a compact cell design. The catholyte is forced through this three-dimensional cathode inward through the bottom, upward between membrane and cathode and outward at the top. The process is covered by the following 10 US Patents, naming D.W. Cawfield as the inventor or co-inventor: US 4,740,287; 4,743,350; 4,761,216; 4,770,756; 4,784,875; 4,793,906; 4,836,903; 4,857,158; 4,892,636 and 4,992,147.

It should be noted that, if a three-dimensional electrode of significant thickness is used with a poorly conducting electrolyte, the voltage drop in the electrolyte filling the voids in the electrode may cause a poor distribution of CD. In such a case it may be desirable to lower the conductivity of the electrode to match that of the electrolyte (compare Section 5.1.11).

2) Hydroxylammonium nitrate can be made by reduction of nitric acid (HAN process, Fig. 5.28). This compound is of interest to the military as a component of liquid monopropellants. The HAN process uses a mercury cathode for high hydrogen over voltage to give 85% CE at the cathode. Seventy percent nitric acid is the feed for the cathode compartment; nitric acid is also the anolyte. The anode is platinum clad niobium. The process is covered by the following US Patents: 4,849,073; 5,213,784; 5,258,104; and 5,510,097.

3) The on-site generation of chlorine dioxide by anodic oxidation of a solution of sodium chlorite (DCD process, Fig. 5.29).

The process uses a zero gap, single pass flow through cell. A high surface area anode structures provides high (90%+) conversion of chlorite to chlorine dioxide in a single pass at ambient temperature, resulting in a high-strength chlorine

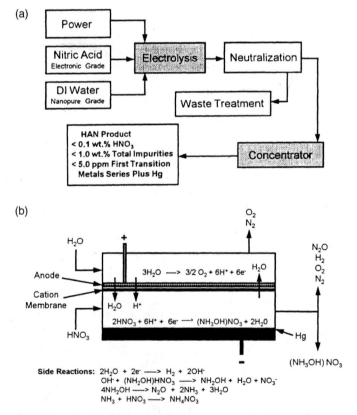

Figure 5.28 Hydroxylammonium nitrate process overview (*courtesy:* The ElectroSynthesis Co. [87]). (a) Overview of HAN process. (b) Chemistry of HAN process electrolysis.

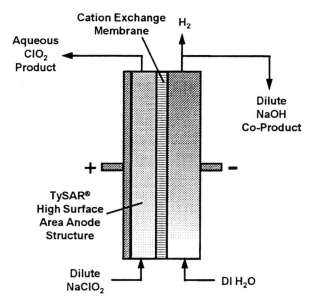

Figure 5.29 Electrochemical cell configuration for on-site generation of chlorine dioxide (*courtesy*: The ElectroSynthesis Co. [87]).

dioxide solution (3–20 g/l). The addition of salts to the anolyte significantly improves CE and solution conductivity.

This process is covered by the following six US Patents in which D.W. Cawfield and J.J. Kaczur have been named as inventors: US 5,041,196; 5,084,149; 5,092,970; 5,106,465; 5,158,658; and 5,294,319.

5.2 Sensors and Actuators

Perfluorinated ionomers are being used in a wide variety of *sensor* applications. A commonly used carbon monoxide (CO) detector for instance uses a Nafion electrolyte [32,33]. Jiang et al. have reported on the use of Nafion in a sensor for both CO and methanol [34]. Stability of the response of these CO sensors has been a problem and has led to the recall of the "Nighthawk" and "Lifesaver" detectors in 1999 [35]. It is believed that it was due to the absorption of organic contaminants from the air or from packaging material into the Nafion electrolyte, where they may undergo acid catalyzed condensation reactions. One approach to overcome this problem is to cover the sensor with a thin layer of Teflon AF, which is quite permeable to CO [36]. In another approach, the Nafion electrolyte is converted to the sodium ion form. Whenever a sensor, using a fluorinated sulfonic ionomer as the electrolyte, is exposed to possibly contaminated gases, the conversion of the ionomer to the sodium (or lithium) form should be considered as an approach to long-term stability. In still another approach, the stability is improved by maintaining the Nafion electrolyte wet [37]. This same publication also describes a manufacturing technology suitable for inexpensive mass production. The AC impedance of a planar CO detector using Nafion as the electrolyte was studied by Mortimer et al. They found that the impedance of commercially available extruded film is lower than that of cast Nafion [38].

The general concept of electrochemical sensor based on a Nafion electrolyte has been studied by Opekar and Stulik [39]. In many cases, these sensors respond to an oxidizing or reducing agent. Do and Chang described an amperometric nitrogen dioxide gas sensor based on polyaniline deposited on a gold/Nafion electrode [40]. The determination of NO is of particular interest for biological systems. Wu et al. reported on the electrocatalytic oxidation of NO on electrodes modified by muti-walled carbon nanotubes. Coating the tubes with a solution of Nafion eliminated the interference by anions such as the nitrite ion and some biological substances [41]. Katrlik and Zalesakova obtained similar results on a bare carbon fiber microelectrode [42]. Joshi et al. used a Nafion-coated platinum disk electrode for the *in vitro* determination of NO during cardiac allograft rejection [43]. Chang et al. described a sensor array for *in vitro* monitoring of NO. The sensor system involved a nickel tetrasulfonated phthalocyanine film electrodeposited on the surface of a carbon electrode and subsequently covered with a layer of Nafion. To demonstrate the potential of this system for drug screening, the sensor was used to monitor the response of cultured A172 glioblastoma cells to chemicals of biomedical relevance [44].

Another important sensor application for biological systems is the determination of glucose. A large number of publications on this subject are available. In most cases, an immobilized enzyme, such as glucose oxidase, is used. In a most recent publication [45], Ghica and Brett further mediated the system with methyl viologen. The enzyme and the mediator were immobilized in Nafion ionomer. A detection limit of 20 μM was observed. Lim et al. described a glucose biosensor made by electro co-deposition of glucose oxidase and palladium nanoparticles onto a NAFION®-coated carbon nanotube film. A top coat of Nafion was used to eliminate the interference by uric or ascorbic acid [46].

The most important application for glucose sensors is of course in the control of diabetes, particularly in combination with an insulin pump, to maintain a constant desirable glucose concentration with minimum intervention by the patient or a doctor. One design of such a sensor is based on two sets of stainless steel microneedles that are about 1.3 mm long. This is long enough to penetrate the stratum corneum and access the interstitial fluid in the subcutaneous tissue, but short enough to avoid contact with the nerve fibers. Mounted on a flexible non-conducting substrate such a sensor can be wrapped around a finger or attached to some other area of skin. One set of microneedles is silver plated and then coated with silver chloride and acts as the reference electrode. The other set is electroplated with rhodium and then coated with a layer of cellulose acetate containing glucose oxidase immobilized by glutaraldehyde. In this layer, glucose is oxidized generating hydrogen peroxide, which can then be sensed at the rhodium electrode. A top coat of Nafion prevents interference by other biological components, such as ascorbic acid.

Yu et al. used Nafion as the electrolyte for an electrochemical sulfur dioxide gas sensor. The stability of the sensor was improved by pretreating the Nafion with sulfuric acid [47].

Schiavon et al. have suggested a very sensitive detector for gaseous hydrogen sulfide. This sensor uses a porous silver electrode supported by a sheet of Nafion 417. The counter- and reference electrodes are in an acidified sodium perchlorate solution on the backside of the membrane. In the amperometric mode of operation, the silver electrode is held at a potential at which it would not release electrons unless such oxidation is enhanced by the formation of highly insoluble silver sulfide, similar to the tarnishing of silver. In this mode, a detection limit of 45 ppb (v/v) is reported. Because the sensing electrode is exposed directly to the gas stream, the response time for 95% of full current is only 0.5 s. The sensor is even more sensitive (down to 0.07 ppb) in the cathodic stripping mode: in this mode, the silver sensing electrode is operated for a certain period of time at constant potential to allow the accumulation of silver sulfide, followed by current reversal and measuring the current necessary for the complete reduction of the accumulated silver sulfide [48].

Not all sensors using fluorinated ionomers are based on electrochemical detection. optical sensors may involve a color forming reagent, which is immobilized in a transparent ionomer film: Amini et al., for instance, immobilized 2-(5-bromo-2-pyridylazo)-5-(diethylamino)phenol in a Nafion film cast on a glass substrate. This sensor is highly transparent, mechanically stable and showed no evidence of reagent leaching. It was used for the spectrophotometric determination of nickel in aqueous solution in a flow-through cell [49]. Sands used 4-decyloxy-2-(2-pyridylazo)-1-naphthol in a very similar system for the detection of copper. The exposure time of the sensor to the solution in a flow-through system was adjusted according to the expected copper concentration: for the analysis of a copper alloy 90 s exposure was used while for the determination of copper in tap water 600 s was used [50]. Capitan-Vallvey used bathocuproine as a chromogenic reagent and *p*-anisidine as a reducing agent in a simple disposable test strip for the determination of copper in humane serum [51].

Actuators are devices that perform a movement in response to an electrical stimulus. The movement can be the result of many different effects, for instance the piezoelectric effect. The present discussion is limited to actuators that utilize an ionomer, which will undergo a deformation as a result of a mass transport caused by an applied electrical current. In a typical application, platinum or gold electrodes are applied to the two surfaces of an ionomer film (ionic polymer metal composites or IPMC). If a current is applied between the electrodes, cations in the ionomer film will move toward the cathode together with some water of hydration and cause the surface near the cathode to expand, resulting in the bending of the film. Shahinpoor has published a review of IPMC actuators [52]. Advantages of IPMCs are the low voltage requirements (about 2 V) and the easy production methods, particularly for large numbers of small parts, which can be cut out of a preformed "MuscleSheet" (™Biomimetic Products, Inc.). The following results are based on 25 mm × 3.5 mm strips cut from such a sheet based on Nafion 117 coated on both sides with platinum [53]: the deflection is a function of the applied voltage and increases from about 1 mm at 1 V to 3 mm at 2 V, 7 mm at 3 V and 12 mm at 4 V. Further increase in voltage results in only a small increase in deflection. A low frequency (1 Hz) AC voltage causes a smaller, oscillating deflection, but the amplitude of this oscillation is maintained almost unchanged over a million cycles. The deflection is smaller at lower temperatures, but even at –140°C a deflection of 2 mm is observed. Using a 1-and-2 V/35 s square wave drive, the actuator draws an initial current of 8 and 20 mA, respectively, for a few seconds; the current then decays to almost zero. When the actuator is short circuited during the off cycle, a reverse current of 7 and 10 mA are discharged for a few seconds.

Actuators based on Nafion sulfonic ionomer will gradually relax after the current is shut off. They will then curve in the opposite direction (toward the cathode) and eventually return to close to the original shape. If the two electrodes are short circuited, the reversal of the curvature is accelerated. Actuators based on Flemion carboxylic ionomer do not exhibit curvature reversal on current interruption [54].

The following explanation is offered for the reversal of curvature: during the current pulse, sodium ions together with water of hydration are pushed toward the cathode, where the sodium ions form sodium hydroxide due to the cathodic decomposition of water. The resulting volume increase, mostly due to the water of hydration, causes the initial curvature toward the anode. At the same time, sodium ions removed from the ionic sites near the anode are replaced by hydrogen ions formed by the anodic decomposition of water. When the current is interrupted, the ionomer near the anode has been, at least partially, converted to the free acid form. The free acid form has higher water absorption than the sodium form and water will be drawn back toward the anode causing the reverse curvature. In the absence of current, neutralization of the acid groups near the anode is retarded by the slow movement of negatively charged hydroxide ions. If the two electrodes are short circuited, movement of the sodium ions alone is sufficient for this neutralization. With a carboxylic polymer, undissociated carboxylic acid groups are formed near the anode. These groups exhibit less water absorption than the corresponding sodium form and no reverse curvature results.

A somewhat different actuator, acting as an acoustic transducer, has been claimed by Waltonen and Schutz [55]. For this application, a film of Nafion is first converted to the sodium, potassium or calcium form and then treated with a solution of a cationic dye for about 25 min. The objective of these treatments is to increase the resistance of the film. Unfortunately, the resistance values and units given in the reference do not make any sense. The surfaces of the film are then coated with a layer of gold or aluminum of about 50 nm thickness. The transducer is particularly useful as a transmitter and receiver for ultrasound for medical applications. As a transmitter, it responds to signals of only 2 or 3 V compared to 60–100 V for typical ceramic piezoelectric devices.

5.3 Dialysis

Ionomer membranes allow several mass transport modes, not involving a net electrical current flowing through the membrane. Depending on whether the fluid on one or both sides of the membrane is in liquid or gas form, one can distinguish between dialysis, pervaporation and gas diffusion. These three modes of mass transport are possible in ionomer membranes as well as in membranes made of uncharged polymers. Donnan dialysis, on the other hand, is a special case of dialysis unique to ionomer membranes. All these processes are driven by differences in concentration (or partial pressure in the case of gases or vapors), rather than by differences in electrical potential or total pressure.

In Donnan dialysis, ions of the same charge move in opposite direction through an ionomer membrane driven by a difference in concentration of one or more species. The rates of transport have to be in balance to maintain electrical neutrality (zero current). However, even though there are no electrodes involved, electrical potential differences are still a factor, as illustrated in the following example [56]: Bandini describes a process for conversion of sodium phenoxide into undissociated phenols, in which a solution of the sodium phenoxide is placed on one side of a Nafion membrane and dilute hydrochloric acid on the other. Both sodium and hydrogen ions can diffuse through the membrane driven by the concentration gradients. Because the hydrogen ion is about 7 times as mobile as the sodium ion, its movement generates an electrical potential difference between the two solutions that retards the hydrogen ions and accelerates the sodium ions until both are in equilibrium. Because phenols are very weak acids, the phenoxide anions act as proton scavengers, maintaining the hydrogen ion concentration on the organic side very low. As long as there is a stoichiometric excess of acid, the removal of sodium ions will continue even if the sodium ion concentration on the acid side is higher than on the organic side.

Compared to direct addition of hydrochloric acid, the membrane process eliminates the separation of the product from by-product sodium chloride (compare salt splitting). The process is applicable in cases where the cations of two compounds should be exchanged without mixing the anions or other components of the system.

Yang and Pintauro [57] developed a transport model for the Donnan dialysis of Cs/Pb and Na/Pb mixtures using Nafion 117 membrane. The model involves cylindrical pores of variable diameter and/or variable wall charge density. Excellent agreement with experimental data was obtained.

In *pervaporation*, the ionomer membrane separates a liquid from a gas or vapor phase. The liquid phase typically consists of a mixture of two or more volatile components, and the objective of the process is to achieve a separation of these components not possible by evaporation only.

One application is the dehydration of nitric acid to break the azeotrop at 68% nitric acid concentration. Sportsman et al. [58] used a commercially available Nafion 90209 sulfonate/carboxylate bilayer membrane for this purpose. Preferential water removal was observed with feed concentrations of up to 80% nitric acid. Maximum water transport rates of 0.2 kg/m^2 h were found at a feed concentration of 50% nitric acid and a permeate pressure of less than 8 mmHg. Ames et al. [59] used a cast film of Nafion carboxylate polymer to improve nitric acid rejection. Compared with a sulfonate film of similar thickness (Nafion 111), water separation efficiency was indeed one order of magnitude higher; however, permeate fluxes were two orders of magnitude smaller. These results were probably affected by partial conversion of the polymer to the ester form during dissolution and casting.

The dehydration of aqueous acetic acid was studied by two groups of authors: Ray et al. [60] compared the performance of Nafion with that of four different acrylonitrile copolymers and a polyimide membrane. Nafion had the highest flux but the poorest selectivity. Polyimides exhibited good selectivity with very low flux. Films made of acrylonitrile copolymers, particularly those using hydroxyethyl methacrylate as a co-monomer, showed reasonable flux and selectivity, Kusumocahyo and Sudoh [61] converted Nafion to an alkyl ammonium ion form to improve selectivity with some penalty in flux. Tetra octyl ammonium ion appeared to be most suitable with a flux of 0.18 kg/m^2 h and a selectivity of 243 for the feed concentration of 90% acetic acid.

The separation of olefins from saturated hydrocarbon using Nafion in the silver ion form was studied in a liquid/liquid, liquid/vapor and vapor/vapor systems. Sungpet et al. reported on the transport of cyclohexene, 1,5-hexadiene, 1-hexene and styrene from a 0.5 M solution in isooctane using a Nafion/poly(pyrrole) composite [62]. TanyaKao et al. [63] reported on the separation of benzene and cyclohexane by pervaporation through a Nafion 117 membrane in the silver ion form and swollen with glycerol. Although with pure solvents, the sorption of benzene was only 1.5 times higher than that of cyclohexane, from the mixture, this ratio was between 15 and 19. Additional examples of the use of Nafion in the silver ion form for the separation saturated from unsaturated hydrocarbons will be provided in Section 5.4.

5.4   Gas and Vapor Diffusion

Gas permeation data for the precursor polymers were not available; data for FEP perfluorinated ethylene–propylene co-poplymer at 25°C are given as an approximation: hydrogen = 14; helium = 40; nitrogen = 2.15; oxygen = 5.9; and carbon dioxide = 1.7 Barrers (1 Barrer = 10^{-11} cm^3 × cm/cm^2 × s × mmHg). For the ionic forms, an enhancement of the transport of water soluble gases, such as carbon dioxide, is observed. for the hydrogen form of Nafion, the following values are reported: hydrogen = 15, nitrogen = 0.5, oxygen = 1.5, carbon dioxide = 300, and carbon monoxide = 30. This is a case of facilitated transport where an additive in the polymer film (in this case water) selectively enhances the transport of some gases [64].

In the case of carbon dioxide, ethylene diamine (EDA) is the additive that has received the most attention [65,66]. When a perfluorinated sulfonic acid ionomer, such as Nafion is equilibrated with an excess of EDA in aqueous solution, the monoprotonated $EDAH^+$ cation will be the counter-ion in the polymer. Depending on the EDA concentration in the solution, the ionomer will imbibe an additional amount of free EDA base as a highly polar solvent. One may expected that in such a medium carbon dioxide would be absorbed in the form of a carbonate or bicarbonate ion. However, Yamaguchi et al. suggested the reaction of carbon dioxide with the free amino group of the $EDAH^+$ cation would result in the formation of a carbamate zwitterion and that this ion is the actual carbon dioxide carrier [67].

Silver ions form adducts with olefins and the separation of olefins from other gases using ionomer membranes in the silver ion form has been the subject of a number of studies. Eriksen et al. [68] reported on the separation of a humidified mixture of ethylene and ethane through a Nafion 117 film in the silver and the sodium ion form. The same authors obtained higher fluxes if the water content of the film was increased by a hot glycerol treatment [69]. Sungpet et al. added 2% poly(pyrrol) to the silver ion form of Nafion to create a mixed ionic/electronic conductor as the appropriate electronic environment for the reaction of silver ions with ethylene [70].

The drying and the humidification of gases are of some importance. Equipment based on capillary tubing made from Nafion sulfonic polymer is commercially available from PermaPure (Figs 5.30–5.33). A schematic of such a tubular humidity exchanger is shown in Fig. 5.31. The drying of gases may be done for analytical purposes, while the humidification of hydrogen is of interest in fuel cells. The dryer may contain a single tube of Nafion (designated MD followed by the tube diameter in mils) or a tube bundle (designated PD followed by the number of 30 mil o.d. tubes in the bundle). The final number in the code indicates the length of the tubes in inches. The code MD-050-24 therefore indicates a dryer consisting of a single tube of 50 mil (=1.27 mm) diameter and 24 in. (=610 mm) length. Any letter after the numbers indicates the material of the housing, for instance S = stainless steel or P = polypropylene.

The performance of the dryers is indicated by the dew point of the exit gas as a function of gas flow. The incoming sample gas is saturated with water vapor at 20°C. The flow of drying gas is twice the flow of sample gas. It can be seen that the performance is proportional to the number of tubes in the bundle, but not necessarily to the length of the tubes.

If the gas to be dried is inexpensive, available at a pressure substantially higher than needed for the dry gas product and partial recovery of this

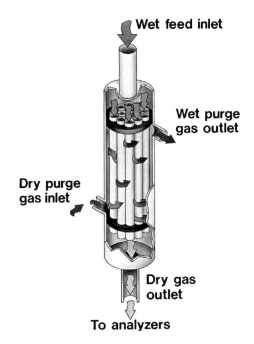

Figure 5.30 Tubular humidity exchanger [88].

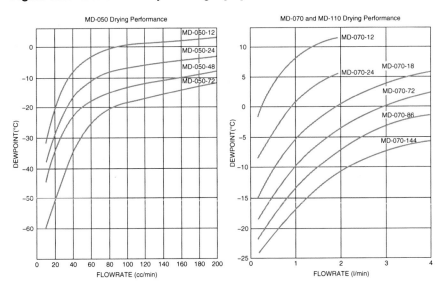

Figure 5.31 Performance of some single tube dryers [88].

gas is acceptable, then a fraction of the dried gas can be used as a purge. This is illustrated by using the generation of dry air from compressed air as an example: compressed air is used as the "Wet Feed In". At the end of the bundle, a let down valve is used to decompress the air to the

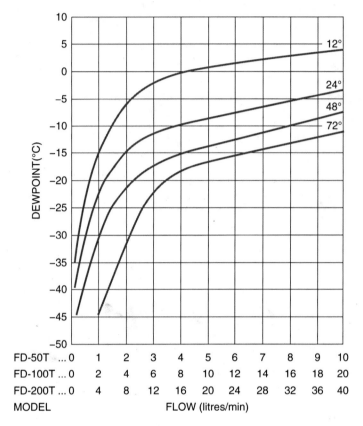

Figure 5.32 Performance of multi-tube dryers [88].

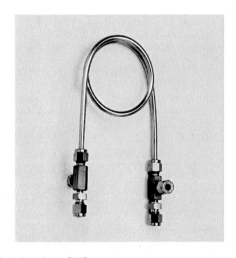

Figure 5.33 Single tube dryer [88].

pressure required for the end use with a corresponding increase in volume. A fraction of this "Dry Gas" product air is then further decompressed to essentially atmospheric pressure and used as the "Dry Purge Gas in". Even though only a fraction of the product is used as the purge, the *volume* of this fraction is larger than the volume of the compressed air in. As a result, even if the purge picks up the entire water content of the wet feed, its partial pressure of water vapor is always lower than that of the compressed feed. This differential in partial pressure drives water through the wall of the tubing. The beauty of this device is its simplicity, with no desiccant that needs regeneration, and long-term silent operation without any electric energy input. The energy need to drive the water movement is instead derived from the energy content in the compressed air. An extensive analysis of the efficiency of membrane dryers can be found in Ref. [71].

5.5 Protective Clothing

Many materials are available that are impervious to harmful agents. However, to be useful for a practical protective clothing, the material must also be able to transmit water vapors at a high rate. This is necessary for the comfort of the wearer, particularly at high ambient temperatures and/ or under physical exertion. In the past, the combination of a breathable, porous material with an adsorbent, such as activated carbon, has been the approach to overcome this problem.

Ionomers have high rates of water vapor permeation and, therefore, provide wearer comfort. Perfluorinated ionomers in addition exhibit excellent chemical stability, even against the most corrosive agents. Their use in protective clothing has been claimed in Ref. [71]. The degree of swelling of the ionomer is of course a factor in its ability to reject toxic agent. Rivin et al. have studied the permeation of dimethyl methylphosphonate (a nerve gas stimulant), water and propanol through Nafion [72].

5.6 Catalysis

Perfluorinated sulfonic acid ionomers have attractive properties for catalytic applications:

1) Very high acid strength (super-acid).
2) Exceptional chemical and thermal stability.

3) The solid state allows easy separation from the products formed; in membrane form this catalyst may actually prevent mixing of the reactants [73].

Three different types of catalysis may be considered:

1) Catalysis by the hydrogen ion.
2) Catalysis by some other cation acting as a counter-ion; some hydrogen ions may also be present.
3) Catalysis by some solid particles, such as metal particles, present in the ionomer.

Acid catalysis using perfluorinated ionomers in the sulfonic acid form has been studied most extensively. Olah has published a comprehensive review of this subject, including 181 references [74]. The high catalytic activity of these polymers is a result of the electron withdrawing effect of the perfluoroalkyl group on the sulfonic acid site. The acid strength of Nafion is comparable to that of 96–100% sulfuric acid. Gas phase alkylations using Nafion result in cleaner reaction products with less by-products formed. The alkylation of benzene with ethylene and propylene are given as examples with experimental procedures. When olefines are used as alkylating agents, the catalytic activity of Nafion slowly decreases, possibly due to deactivation of catalytic surface sites by some polymeric material. If alcohols are used as alkylating agents, the lifetime of the Nafion catalyst is actually improved; however, ethers may be formed as by-products.

Alkylations, disproportionations and transalkylations of alkylbenzenes typically require temperatures in excess of 160°C and the long-term stability of an ionomeric catalyst is of concern under these conditions. In the isomerization and disproportionation of m-xylene, Nafion suffered a 10% loss of activity after 100 h of operation. A 2–3% weight gain and a 5% loss in titrable acidity were observed.

Aromatic acylations using aroyl halides require only 10–30% of the stoichiometric amount of Nafion catalyst in contrast to the typical inorganic Friedel–Crafts catalysts, which usually require more than molar equivalence of catalyst. Attempted acylation of aromatics with acetyl chloride, on the other hand, resulted in the formation of ketene and diketene.

Nafion is an effective nitration catalyst. In many cases, the product water can be removed by azeotropic distillation. This eliminates the need for the recovery of spent sulfuric acid as required in conventional nitrations.

In the above-mentioned examples, the perfluorinated ionomer has been used in the free sulfonic acid form; there are other cases where a catalytically active metal ion occupies all or part of the ion exchange sites. Nafion in which part of the hydrogen ions has been replaced by mercury ions yields more of the meta-isomer when used as a catalyst in the nitration of toluene, ethyl benzene and *tert*-butyl benzene. In the case of ethyl benzene, side chain oxidation is also suppressed.

A partially (about 25%) mercury-exchanged Nafion is also an effective catalyst for the hydration of alkynes to yield carbonyl compounds. The older technology, using mercuric salts dissolved in a liquid acid, results in the formation of a by-product of an inactive sludge containing dispersed mercury and organic mercury compounds. The improper disposal of this waste product was the cause of the Minamata disaster (see Chapter 1). The use of mercury ions confined inside a solid acid avoids this problem. One could say that Nafion has come full circle.

Publications in "Synthesis" relating to the use of Nafion as a catalyst

Subject	Year	Page
Acylation of aromatics using aroyl chlorides or anhydrides (e.g., benzophenones)	1978	672
Nitration of aromatic compounds using n-butyl nitrate	1978	690
Formation of methoxy methyl ethers using dimethoxy methane	1981	471
Cleavage of these ethers	1983	892
Tetrahydropyranylation of alcohols using dihydro-4H-pyran (and cleavage)	1983	894
O-trimethylsilylation of alcohols, phenols and carboxylic acids	1983	894
Ring closure of diols	1981	474
Dimethyl acetals from aldehydes or ketones using trimethyl orthoformate	1981	282
Hydrolysis of benzophenone dimethyl acetal	1981	283
Ethylene dithioacetals	1981	283
1,1-Diacetates from aldehydes using acetic anhydride (at ambient temperature!)	1982	963
Hydration and methanolysis of epoxides	1981	281
Rupe rearrangement of alkenyl tertiary alcohols to olefinic ketones	1981	473
Hydration of alkynes (Nafion in the mixed H^+/Hg^{++} form)	1978	671

The trimethylsilyl ester of Nafion has been prepared by heating the sulfonic acid resin with chloro trimethylsilane and a drop of concentrated sulfuric acid to 80°C for 5 h. A variety of compounds possessing active hydrogen atoms can be silylated by this reagent [75]. The reagent is available from Aldrich.

Waller has published two review articles on the use of Nafion, both in the hydrogen and a metal ion form, as a catalyst [76,77]. More recently, the subject has been covered by publications by Seen [78], Harmer and Sun [79] and Schneider et al. [80].

The photo-assisted oxidation of pollutants was studied by several researchers using ferrous ion immobilized on a Nafion membrane. Chlorinated phenols and the azo dye Orange II were used as model pollutants. Parra et al. used a solution cast Nafion membrane in which the ferrous ion was directly introduced into the casting solution [81]. Superior performance compared to Fe-exchanged Nafion 117 was reported.

References

1. Bianchi, G., de Nora, V., Gallone, P., Nidola, A., US Patent 3,616,445 assigned to Electronor Corp. 1967.
2. de Nora, V., Kuehn von Burgsdorff, J.-W., Chem. Ing. Tech., **47**, 125–128, 1975.
3. Trasatti, S., O'Grady, W.E., Advances in Electrochemistry and Electrochemical Engineering, Vol. 12, J. Wiley & Sons, New York, 1981, pp. 177–261.
4. Kasuya, K., US Patent 4,465,580 assigned to Chlorine Engineers Corp., 1984.
5. Twardowski, Z., US Patent 5,587,083 assigned to Chemetics International, Dec. 24, 1996.
6. Brown, P., US Patent 4,036,713 assigned to Foote Mineral Co., Jul. 19, 1977.
7. Wade, R., Guilbault, L., US Patent 4,394,226 assigned to Thiokol, Jul. 19, 1983.
8. Sharifan, H., Tanner, R., US Patent 4,917,781 assigned to Southwestern Analytical Chemicals, Apr. 17, 1990.
9. Aoyama, T., Shima, E., Ishikawa, J., Sakurai, N., US Patent 4,776,929 assigned to Mitsubishi Gas Chemical Co., Oct. 11, 1988.
10. Shimizu, S., US Patent 4,572,769 assigned to Tama Chemicals Co. Feb. 25, 1986.
11. Klotz, H., Weber, R., Lonhoff, N., Block, H., Pinter, H., US Patent 5,127,999 assigned to Bayer AG, Jul. 7, 1992.
12. Ahmed, M., Chang, H., Seman, J., Holsen, T., J. Membr. Sci., **197**(1–2), 63–74, Mar. 2002.
13. Hillebrand, E.-W., US Patent 6,602,394 assigned to Walter Hillebrand GmbH, Aug. 5, 2003.

14. Dong, D., Mumby, T., Jackson, J., Rogers, D., US Patent 5,643,437 assigned to Huron Tech, Jul. 1, 1997.
15. Fleming, C., Grot, W., Thorpe, J., US Patent 5,411,575 assigned to DuPont, May 2, 1995.
16. Takenaka, H., Torikai, E., Kawami, Y., Wakabayashi, N., Int. J. Hydr. Energ., 7(5), 397–403, 1982.
17. Michas, A., Millet, P., J. Membr. Sci., 61, 157–165, 1991.
18. Holze, R., Ahn, J., J. Membr. Sci., 73(1), 87–97, Oct. 1992.
19. Ioroi, T., Yasuda, K., Siroma, Z., Fujiwara, N., Miyazaki, Y., J. Pow. Sourc., 112(2), 583–587, 2002.
20. Menth, A., Muller, R., Stucki, S., US Patent 4,312,736 assigned to Brown Boveri, Jan. 26, 1982.
21. Han, S.D., Kim, J.D., Singh, K.C., Chaudhary, R.S., Indian J. Chem. Section A – Inorg. Bio-Inorg. Phy. Theor. Anal. Chem., 43(8), 1599–1644, 2004.
22. Stucki, S., Kotz, R., J. Electro-analytical Chem., 228, 407–415, 1987.
23. Baumann, H., Setz, W., Swiss Biotech., 6, 16–18, 1988.
24. Stucki, S. et al., J. Electrochem. Soc., 132, 367, Feb. 1985; also, J. Appl. El. chem., 17, 773, 1987.
25. Menth, A., Stucki, S., US Patent 4,416,747 assigned to Brown Bovery, Nov. 22, 1983.
26. Andrews, C., Murphy, O., US Patents 6,712,951, 6,746,580 and 6,866,806 assigned to Lynntech International, Mar. 30 and Jun. 8, 2004 and Mar. 15, 2005.
27. Onda, K., Ohba, T., Kusonoki, H., Takezawa, S., Sunakawa, D., Araki, T., J. El.chem. Soc., 152(10), D177–D183, Oct. 2005.
28. Stucki, S., US Patent 4,455,203 assigned to Brown Bovery, Jun. 19, 1984.
29. Ando, Y., Tanaka, T., Inter. J. Hyd. Ener., 29(13), 1349–1354, Oct. 2004.
30. Millauer, H., US Patent 4,014,762 assigned to Hoechst AG, Mar. 29, 1977.
31. Millauer, H., Chem. Ing. Tech., 52(1), 53–55, 1980.
32. Shen, Y., Consardoru, F., Field, G., US Patent 5,573,648 assigned to Atwood Systems, Nov. 12, 1996.
33. Prohaska, O., LaConti, A., Giner, J., Manoukian, M., US Patent 6,682,638 assigned to Perkin Elmer LLC, Jan. 27, 2004.
34. Jiang, J., Kucernak, A., J. Electroanal. Chem., 576(2), 223–236, Mar. 1, 2005.
35. The Washington Post, Mar. 19, 1999.
36. Yasuda, A., Yamaga, N., Doi, K., Fujioka, T., Sensors and Actuators B: Chem., 21(3), 229–236, 1994.
37. van der Wal, P., de Rooij, N., Koudel-Hep, M., Analusis, 27(4), 347–351, 1999; also, Sensors and Actuators B: Chem., 35(1–3), 119–123, Sep. 1996.
38. Mortimer, R., Beech, A., Electrochim. Acta, 47(20), 3383–3387, Aug. 5, 2002.
39. Opekar, F., Stulik, K., Anal. Chim. Acta, 385(1–3), 151–162, Apr. 5, 1999.
40. Do, J.-S., Chang, W.-B., Sensors and Actuators B: Chem., 101(1–2), 97–106, Jun. 15, 2004.
41. Wu, F., Zhao, G., Wei, X., Electrochem. Commun., 4(9), 690–694, Sep. 2002.
42. Katrlik, J., Zalesakova, P., Bioelectrochem., 56(1–2), 73–76, May 15, 2002.

43. Joshi, M., Lancaster, J., Liu, X., Ferguson, T., Nit. Oxide-Biol. Chem., **5**(6), 561–565, 2002.
44. Chang, S.-C., Cole, A., Bedioui, F. et al., Biosensors & Bioelectronics, in press.
45. Ghica, M., Brett, M., Anal. Chem. Acta, **532**(2), 145–151, Mar. 15, 2005.
46. Lim, S.H., Wei, Ji, Lin, J., Li, Q., You, J.K., Biosen. Bioelectr., **20**(11), 2341–2346, May 2005.
47. Yu, C., Wang, Y., Hua, K., Xing, W., Yang, H., Lu, Chinese J. Anal. Chem., **30**(4), 397–400, 2002.
48. Schiavon, G., Zotti, G., Toniolo, R., Bontempelli, G., Anal. Chem., **67**(2), 318–323, 1995.
49. Amini, M., Momeni-Isfahani, T., Khorasani, H., Pourhossein, M., Talanta **63**(3), 713–720, 2004.
50. Sands, T.J., Cardwell, T.J., Cattrall, R.W., Farrell, J.R., Iles, P.J., Kolev, S.D., Sensors and Actuators B: Chem., **85**(1–2), 33–41, 2002.
51. Capitan-Vallvey, L.F. Cano-Raya, C., del Valle, C., Ramos, M., de Orbe-Paya, I., Avidad, R., Gonzales, V., Anal. Lett., **35**(4), 615–633, 2002.
52. Shahinpoor, M., Electrochim. Acta, **48**(14–16), 2343–2353, Jun. 30, 2003.
53. Bar-Cohen, Y., Xue, T., Shahinpoor, M., Simpson, J., Smith, J., Proceedings SPIE's 5th Ann. Intn. Symp. Smart Structures and Materials, Mar. 1–5, 1998, San Diego, CA, Paper 3324-32.
54. Le Guilly, M., Xu, C., Cheng, V., Taya, M. et al., Proceedings SPIE, **5051**, 362–371, 2003.
55. Waltonen, J.R., Schutz, R.W., US Patent 5,230,921 assigned to Blacktoe Medical, Jul. 27, 1993.
56. Bandini, S., J. Membr. Sci. **207**(2), 209–225, 2002.
57. Yang, Y., Pintauro, p.n., Ind. Eng. Chem. Res., **43**(12), 297–2965, 2004.
58. Sportsman, K.S., Way, J.D., Chen, W., Pez, G. Laciak, D., J. Membr. Sci., **203**(1–2), 155–166, Jun. 2002.
59. Ames, R., Way, J.D., Bluhm, E., J. Membr. Sci., **249**(1–2), Mar. 2005.
60. Ray, S., Sawant, S., Joshi, J., Pangarkar, V., J. Membr. Sci., **138**(1), 1–17, Jan. 1998.
61. Kusumocahyo, S., Sudoh, M., J. Membr. Sci., **161**(1–2), 77–83, Aug. 1999.
62. Sungpet, A., Way, J., Koval, C., Eberhart, M., J. Membr. Sci., **189**(2), 271–279, Aug. 2001.
63. TanyaKao, S., Wang, F., Lue, S., Desalination, **149**(1–3), 35–40, Sep.10, 2002.
64. Fabiani, C., Teghil, R., Scibona, G., Gas Sep. Purific., **4**(4), 182–184, Dec. 1990.
65. Way, D., Noble, R., Reed, D., Ginley, G., AIChE J., **33**(3), 480–487, Mar. 1987.
66. Way, D., Noble, R., J. Membr. Sci., **46**, 309–324 1989.
67. Yamaguchi, T., Koval, C., Noble, R., Bowman, C., Chem. Eng. Sci., **51**(21), 4781–4789, Nov. 1996.
68. Eriksen, O., Aksnes, E., Dahl, I., J. Membr. Sci., **85**(1), 89–97, Oct. 1993.
69. Eriksen, O., Aksnes, E., Dahl, I., J. Membr. Sci., **85**(1), 99–106.
70. Sungpet, A., Way, J., Thoen, P., Dorgan, J., J. Membr. Sci., **136**(1–2), 111–120, Dec. 1997.

71. Grot, W., Rivers, J., Silva, R., US Patents 4,518,650 and 4,469,991 assigned to DuPont, May 21, 1985 and Sep. 4, 1984.
72. Rivin, D., Meermeier, G., Schneider, N., Vishnyakov, A., Neimark, A., J. Phy. Chem. B, **108**(26), 8900–8909, 2004.
73. Vaughan, R., US Patent 3,976,704 assigned to Varen Technologies, Aug. 24, 1976.
74. Olah, G.A., Iyer, P.S., Prakash, S., Synthesis, Jul. 1986, 513–531.
75. Murata, S., Noyori, R., Tetrahed. Lett., **1980**(21), 767.
76. Waller, F., J. Brit. Poly. J., **16**, 239–242, Dec. 1984.
77. Waller, F.J., VanScoyoc, R.W., Chemtech, **1987**, 438–441.
78. Seen, A. J., J. Mol. Catal. A, **177**(1), 105–112, Dec. 2001.
79. Harmer, M., Sun, Q., Appl. Catal. A, **221**(1–2), 45–62, Nov. 2001.
80. Schneider, M., Zimmermann, K., Aquino, F., Bonrath, W., Appl. Cat. A, **220**(1–2), 51–58, Oct. 2001.
81. Parra, S. et al. Langmuir, **20**(13), 5621–5629, 2004.
82. 11th Krupp Uhde Chlorine Symposium, Dortmund, May 2001.
83. 12th Krupp Uhde Chlorine Symposium, Dortmund, May 2004.
84. www.EUROchlor.com
85. DuPont NAFION® Technical Bulletin 94-03.
86. Grot, W., Paper presented at the ECS Meeting Honolulu, May 1987.
87. Kaczur, J., Paper presented at the 11th Intern. Forum, ElectroSynthesis Co., Clearwater Beach, FL 1997.
88. Permapure, Inc., Product Literature and personal communication, Sep. 2005.

6   Fuel Cells and Batteries

6.1   Introduction

Fuel cells and batteries are electrochemical devices that convert chemical energy directly into electrical energy without combustion. In an internal or external combustion engine, electrons are transferred from the fuel directly to the oxidizer, releasing the stored energy as heat. Converting this heat into mechanical energy is subject to the limitations of the Carnot cycle.

In a fuel cell or battery, the fuel releases electrons to an electrode (the anode). These electrons then travel through an external circuit (or load), doing useful work, until they reach the cathode of the fuel cell or battery, where they combine with the oxidizer. The movement of some ions through the electrolyte that connects the anode and the cathode completes the electrical circuit.

In batteries, the fuel, as well as the oxidizer, is stored inside the battery. When all the fuel is consumed, the battery has to be recharged or replaced. In a fuel cell, the fuel and the oxidizer are stored external to the cell. Theoretically, the fuel cell can, therefore, operate indefinitely as long as fuel and oxidizer are supplied.

There are a number of different types of fuel cells based on the electrolyte used, which in turn determines the operating temperature. The current discussion is limited to fuel cells using a fluorinated ionomer as the electrolyte (PEM fuel cells), with a typical operating temperature of 60–120°C. However, other electrolytes should be briefly mentioned, in order of increasing operating temperature (given in parenthesis): Phosphoric acid (160–220°C), molten carbonate (600–800°C) and solid oxides (800–1000°C). All these can use air as the oxidizer. A special case is aqueous potassium hydroxide (20–100°C), which has many desirable properties, but requires oxygen as an oxidizer because the carbon dioxide content of air would eventually contaminate the electrolyte. This "alkaline" fuel cell is, therefore, used primarily in military and space applications. Note that these different fuel cells do not compete with each other, but each type with other means of generating power. The PEM fuel cells compete with batteries, small internal combustion engines and solar and wind power.

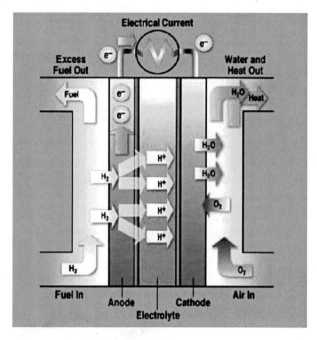

Figure 6.1 Principle of a PEM fuel cell [3].

Figure 6.1 illustrates the principle of a PEM fuel cell: Under the influence of a catalyst (platinum) at the anode, hydrogen gas is dissociated in an equilibrium reaction into protons and electrons. Under zero-current conditions, the equilibrium potential for this reaction is chosen as a reference point for the electrochemical potential series. The equilibrium is not static, but dynamic; that is, the rate of hydrogen dissociation is equal to the rate of recombination of protons and electrons. This rate is referred to as the "exchange current" and it is a function of the catalyst activity and surface area. As a load is applied to the cell, that means electrons are withdrawn from the anode through the external circuit, the anode potential against the electrolyte becomes more positive, the rate of hydrogen dissociation increases while the rate of the reverse reaction decreases. The result is a net dissociation of hydrogen.

As the electrons flow through the external circuit toward the cathode, protons generated by the dissociation of hydrogen flow through the electrolyte to the cathode. At the cathode, protons, electrons and oxygen combine to form water. At zero current, the theoretical potential of the cathode is 1.23 V positive relative to that of the anode, if oxygen is used as the oxidizer and both oxygen and the hydrogen fuel are used at atmospheric pressure. With air at atmospheric pressure as an oxidizer, this potential is slightly lower and it is further reduced as current is drawn.

To maximize the surface area of the expensive catalyst, the catalyst is deposited on the surface of carbon black and distributed in a three-dimensional gas-diffusion electrode (GDE). This GDE, sometimes also called the catalyst layer, may be only a few microns thick, but still has an internal surface area many times larger than the geometric area of the electrode. Carbon black not only provides a high surface support for the platinum catalyst, but also a continuous electronically conducting phase. The ability of carbon black to form long strings of individual particles makes it uniquely suitable for this job. In addition to the continuous electronic conductor, the GDE must also provide a continuous pathway for the ions generated at the anode to reach the ionomer film (the electrolyte) and from there to reach the cathode. Furthermore, there must be a pathway for the gases involved in the electrode reactions. The GDE, therefore, must contain three networks of intermeshing continuous phases, and a platinum particle can act as a catalyst only if it is in contact with all these phases. Figure 6.2 illustrates the placement of platinum particles in this network of three continuous phases.

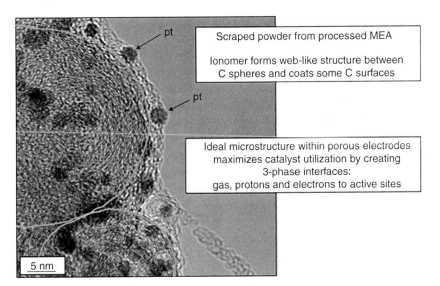

Figure 6.2 Microstructure of a platinum on carbon black catalyst [4].

Looking at the distribution of these three phases along the thickness dimension of the electrode, one can realize that near the boundary with the membrane the current in the catalyst layer is carried mostly by ions. Ionic conductivity is therefore most important in that region. Near the other boundary, the one with the gas-diffusion medium, electronic conductivity and gas permeability are most important. Based on this Xie et al. have constructed graded catalyst layers containing more Nafion® near the boundary with the membrane and observed improved performance compared with catalyst layers of uniform Nafion contend or with layers in which the gradient was reversed [1].

A gradient with respect to pore size and hydrophobicity will be discussed under Section 6.5 (ELAT).

The GDE is made very thin not only to minimize the use of platinum, but also to shorten the pathway for gas diffusion and electronic and ionic conductions. To utilize the short electronic pathway, the current must not be collected at the edge of the GDE, as indicated in Fig 6.2, but from the backside of the GDE, as shown in Fig. 6.3. Between the catalyst layers, acting as electrodes, and the current collector/gas flow field, there are diffusion media to provide current distribution to those parts of the catalyst layer that are above a gas channel (the groves), and gas distribution to areas blocked by the ridges. In a typical cell, a felt or a paper made of

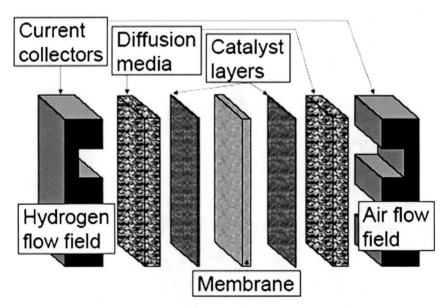

Figure 6.3 Exploded view of a PEM fuel cell [5].

carbon fibers is used as the gas-diffusion medium, and the current collectors are made of graphite. The two catalyst layers are usually bonded or laminated to the central membrane, creating a unit referred to as the "membrane electrode assembly" (MEA).

As stated before, the theoretical cell voltage under no-load conditions is approximately 1.2 V. In a practical fuel cell, the measured open circuit voltage (OCV) is slightly lower, or about 1 V. This is due to a number of loss mechanisms, such as the diffusion of hydrogen gas through the ionomer film to the cathode, where it lowers the cathode potential. As a very small load (0.1 A/cm^2) is applied, the cell voltage drops to 0.8–0.9 V as a result of an increase in electrode overvoltages (Fig. 6.4). Under these conditions, the efficiency of converting chemical into electrical energy is very high (67–75%), but the power output is too low for practical applications. As the load increases to practical levels (about 1 A/cm^2), the cell voltage drops further to 0.5–0.7 V, corresponding to an energy efficiency of about 50%. These additional voltage losses are mostly due to ohmic losses within the cell. While the overvoltage losses are logarithmic functions of current density (CD), the ohmic losses are linear function. A further increase in load will result in progressively larger voltage losses due to mass transport limitations.

A single fuel cell will then deliver a fairly large current (400 A for a 20 cm × 20 cm geometric area) at low voltage. Fuel cells are, therefore, almost always combined in a bipolar stack consisting of many individual cells. The current collectors/flow fields are constructed as bipolar plates, containing gas distribution channels for hydrogen on one side, and for air on the other. Or, particularly in larger fuel cells, the backside of the hydrogen flow field may instead contain channels for a cooling liquid. This side of the plate is then in electrical contact with the cooling surface of the adjacent air flow plate. This geometry maximizes the advantages of thin components: It allows for compact design, minimizes the cost of expensive materials, such as catalysts and ionomers, and reduces the internal electrical and mass transport resistance. The bipolar plates are frequently machined from graphite; more recently, bipolar plates made from conductive polymers have been offered (see Appendix D). Such plates can be molded.

At the endplates, uniform clamping pressure is applied to the stack, not only to seal the individual components, but also to minimize the contact resistance between the current collectors, the diffusion media and the catalyst layers. These three components rely on carbon black as the electronic conductor and the resistance of a carbon–carbon contact depends very much on the contact pressure. Some fuel cells use a pressurized gas bladder to distribute the clamping pressure uniformly over the entire cross-section of the stack. The contact resistance between the catalyst layers and the membrane is minimized by laminating these three components to form an MEA.

6.2 Operating Parameters

The following parameters determine the performance of the fuel cell:

1) Gas pressure and purity: Increased gas pressure results in a small increase in the electrode potentials and improved mass transport. Presence of an inert diluent, such as nitrogen, reduces the partial pressure of the reactant gas and therefore the electrode potentials. It is also a substantial impediment to mass transport. Some impurities, such as carbon monoxide, may act as catalyst poison or diminish the conductivity of the electrolyte (ammonia). A temporary or permanent loss of performance will result. A mixed Pt/ Ru anode catalyst is more resistant to carbon monoxide poisoning than Pt alone.

2) Gas stoichiometry: The stoichiometry is the ratio of the feed rate of reactant gases to the theoretical amount required at the given CD. This parameter is particular important if the reactant gases contain an inert diluent. For instance, if air is used at 2x stoichometry, the part of the fuel cell near the air exit would operate with an oxidant stream containing only about 10% oxygen. Since all parts of the cell must operate at the same voltage, the part near the exit would operate at a lower CD than the rest of the cell. Any back mixing within the gas channels would extend the part of the cell affected. If the CD is changed, gas feeds must be adjusted to maintain the same stoichiometry. Polarization curves should state whether they are run at constant stoichiometry or constant gas flow.

3) Gas humidification: Water is produced at the cathode due to the reaction of oxygen with the electrons and protons transported to the anode by the electronic and ionic conductors, respectively. In addition, protons carry water of hydration to the cathode (electroosmotic water drag). If the hydrogen feed is not sufficiently humidified, this electroosmotic drag may result in drying of the ionomer near the anode with a resulting loss of conductivity. To maintain water

balance inside the ionomer film, a gradient of water content is established within the film that creates a diffusive driving force to make up the difference between humidity feed and electroosmotic drag. With a very thin film, this gradient can be steep enough to eliminate the need for hydrogen humidification.

4) CD and changes in CD.
5) Temperature: While the theoretical voltage is slightly reduced with increasing temperature, the improved conductivity and mass transport result in improved performance up to about 70°C. Above this temperature, problems in water management limit the performance of many currently available ionomers. Operation at higher temperatures, particularly above 120°C, remains an important goal of polymer research.

The parameters listed above are given or set by the operator. The resulting cell voltage and its decline with time and cycling (polarization curves, Fig. 6.4) then describe the performance of the cell or the ionomer used.

As an example of a small stationary fuel cell system, the 5-kW Gen-Core® system by Plug Power, used to provide backup power to telecommunications systems, is chosen. Table 6.1 shows the product characteristics of this system.

To allow operation at temperatures above 80°C, particularly when combined with low humidity conditions, composites of perfluorinated ionomers with inorganic oxides have been studied. Srinivasan et al. compared Nafion 115 with a composite of this polymer with 25% zirconium phosphate.

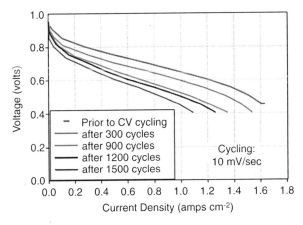

Figure 6.4 Polarization curve before and after 300–1500 cycles [7].

Table 6.1 Product Characteristics of a Commercial 5-kW Fuel Cell Stack [6]

Product Characteristics		5T48	5T24
Performance	Rated net output[1]	0 to 5,000 W	0 to 5,000 W
	Adjustable Voltage	−46 to −56 Vdc (−48)	+25 to +27 Vdc (+24)
	Operating Voltage Range	−42 to −60 Vdc	+21 to +27 Vdc
	Operating Current Range (net)	0 to 109 Amps	0 to 218 amps
Fuel	Gaseous Hydrogen	99.95% Dry	99.95% Dry
	Supply Pressure	80 +/− 16 psig (5.5 +/− 1.1 bar)	80 +/− 16 psig (5.5 +/− 1.1 bar)
	Fuel Consumption	40 standard liters per minute at 3,000W	40 standard liters per minute at 3,000W
		75 standard liters per minute at 5,000W	75 standard liters per minute at 5,000W
Operation	Ambient Temperature	−40°C to 46°C	−40°C to 46°C
	Relative Humidity	0% to 95% Non condensing	0% to 95% Non condensing
	Altitude	−197 ft to 6,000 ft (−60 m to 1829 m)	−197 ft to 6,000 ft (−60 m to 1829 m)
Physical[2]	Dimensions	44' H × 26' W × 24' D (112 cm × 66 cm × 61cm)	44' H × 26' W × 24' D (112 cm × 66 cm × 61 cm)
	Weight	500 Lbs (227 kg)	500 Lbs (227 kg)

		FCC Class A	FCC Class B
Safety	Compliance	FCC Class A UL Listed to ANSI Z21.83 GR-63, GR-1089, GR-487 (NEBS Level 3)[3] CE Certified	FCC Class B UL Listed to ANSI Z21.83 GR-63, GR-1089, GR-487 (NEBS Level 3)[3] CE Certified
Emissions	Water	Maximum 2.0 Liters per hour	Maximum 2.0 Liters per hour
	CO, CO_2, NOx, SO_2	<1ppm	<1ppm
	Audible noise	60 dBA @ 1m	60 dBA @ 1m
Sensors[4]	Gas Hazard Detection	Included	Included
Control	Microprocessor	Included	Included
	2 LED Panel	Included	Included
	Low Fuel Alarm	Included	Included
	Communications[5]	RS-232C Digital Form C Contacts	RS-232C Digital Form C Contacts

[1]Output rated from −40°C to 42°C to 46°C, output decreases 2.5% per degree Celsius. Above 1,000 feet (305 meters), an additional de-rating of 1.5% per 1,000 feet applies.
[2]Excludes fuel storage.
[3]Compliant where applicable.
[4]Optional sensors are available to detect Pad shear, water intrusion and tampering.
[5]Optional communications include MODEM.
Specifications subject to change without notice.

The composite exhibited higher water uptake but slightly lower proton conductivity than the Nafion 115 control. Under low humidity conditions, the performance of the composite in a fuel cell is better than that of Nafion 115 [2]. In an earlier publication [9], the authors described the impregnation of Nafion 115 or a cast film of Nafion with zirconium phosphate by first exposing the film to a solution of zirconyl chloride followed by immersion in 1 M phosphoric acid at 80°C. The resulting composite gave a performance at 130°C and 3 bar hydrogen/oxygen of 1000 mA/cm^2 for the Nafion 115 based composite and 1500 mA/cm^2 for the solution based material, both at 0.45 V. An unmodified Nafion 115 film gave only 250 mA/cm^2 under the same conditions.

6.3 Ionomer Stability

Early in the development of fuel cells using Nafion as an electrolyte, the presence of fluoride ions in the product water was observed. It was concluded that this polymer instability is a result of carboxylic acid end groups terminating the polymer backbone. These end groups are introduced through the use of peroxide initiators. A similar problem had been observed earlier with other perfluorinated polymers, such as FEP and PFA. In the presence of oxygen, degradation of the carboxylic end group results in the conversion of the adjacent carbon atom to another carboxylic acid with the release of two fluoride ions. "Unzipping" of the polymer backbone will therefore continue. The carboxylic end groups can be stabilized by a post treatment of the precursor polymer; for instance with elemental fluorine or a fluorine/nitrogen mixture. While this post-treatment resulted in a very substantial reduction in fluoride ion release, there are still reports of trace amount of fluoride ions released to the product water. The source of this release is not certain at this time: Insufficient fluorine treatment, fluoride ions retained from the hydrolysis step or degradation of the polymer even in the absence of unstable groups.

Many studies have been performed to find correlations between operating conditions on the one hand and decline of performance, release of fluoride ions and formation of carboxylate groups on the other.

It is believed that the degradation is caused by hydroxyl radicals or other reactive intermediates of the four-electron reduction of oxygen at the cathode. However, if an air bleed into the hydrogen stream is used to overcome the poisoning effect of a carbon monoxide contaminant on the anode catalyst, some fluoride release occurs also at the anode (Fig. 6.5). The combination of a carbon monoxide contamination of the hydrogen

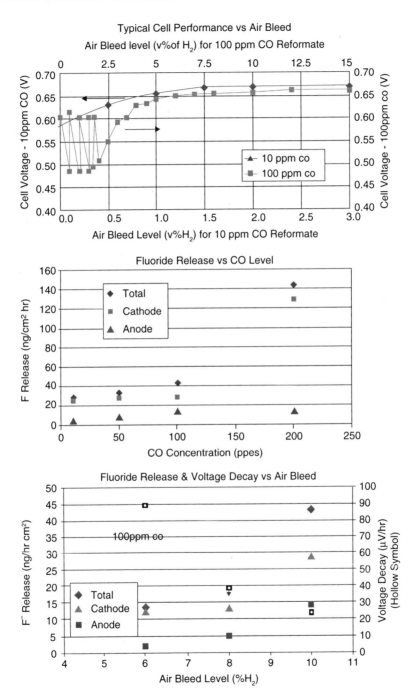

Figure 6.5 System Studies—CO/air bleed and their effect on fluoride ion release [8].

feed with an air bleed to minimize catalyst poisoning actually results in a much larger fluoride ion release than an air bleed alone.

The oxidative attack at the cathode appears to be most severe under open-circuit conditions or with repeated cycling between load and open circuit, because the oxidation potential at the cathode is highest at open circuit. However, Minoru Inaba et al. suggested in a recent publication (*Electrochemical Acta*, in print) that the degradation observed under open circuit occurs at the anode due to oxygen crossover followed by catalytic reaction with the hydrogen.

To study the causes of membrane degradation, chemical tests have been devised to simulate and accelerate the degradation reactions observed in a fuel cell. One such test uses the Fenton's reaction as a source of hydroxyl radicals (compare Section 9.8: Fenton's test). In this test, the sample is exposed to the combined effect of hydrogen peroxide and ferrous ions. To the extent that this redox reaction occurs in the liquid phase, short-lived reaction products may decompose before they can diffuse into the polymer phase, resulting in an inefficient decomposition of hydrogen peroxide. It may, therefore, be better to lock the ferrous ion as a counter ion into the polymer to assure that the decomposition of hydrogen peroxide occurs mostly inside the polymer.

Hicks exposed films made from the 3M polymer as well as model compounds to Fenton's reagent [8]. The model compounds indicated that the following classes of compounds are stable (Fig. 6.6): Perfluorinated sulfonic acids, including those containing ether linkages (MC 5, 7 and 8); otherwise, perfluorinated compounds with a terminal hydrogen atom (MC 6; however, this compound may have been protected by its very low solubility in the reaction medium).

Carboxylic acids are unstable (MC 4); this instability is enhanced by the combination of an ether linkage and a trifluoro methyl group on the carbon adjacent to the carboxylic acid (MC 1 and 2). MC 2 may indicate the effect of solubility on this test. When samples of the 3M polymer were exposed to Fenton's test, Hicks found that loss of mechanical properties is only observed after more than 25% of the mass of the polymer sample is lost. This suggests that scission of the backbone was not a major factor but that unzipping of unstable end groups had occurred (Fig. 6.7).

Otsuka et al. introduced benzene into a hydrogen/oxygen fuel cell system using a Nafion separator and an acidic catholyte containing ferric or cupric ions. The benzene was selectively oxidized to phenol and the formation of ferrous or cuprous ion together with hydrogen peroxide was observed [10]. A type of Fenton's reaction therefore appears likely.

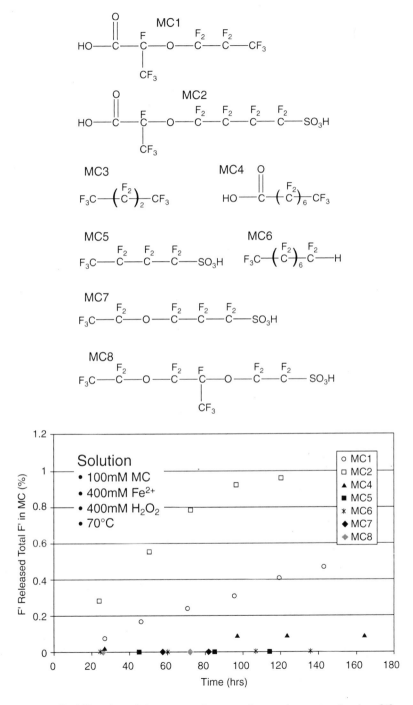

Figure 6.6 Stability of model compounds—membrane decay mechanism [8].

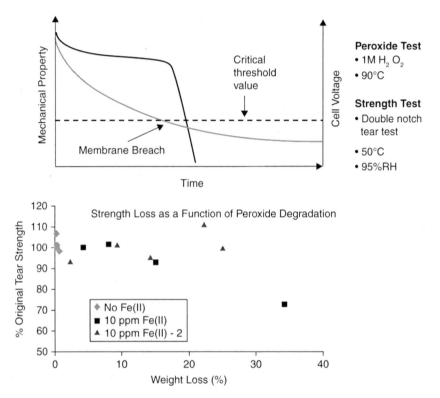

Figure 6.7 Loss of mechanical properties as a result of membrane degradation [8]. A method of aging membranes in a way that degrades mechanical properties is under development.

The term "Fenton's reagent" usually indicates the use of ferrous ions as the single electron reducing agent. However, other single electron reducing agents, such as the titaneous (+3) ion can be combined with hydrogen peroxide. Bosnjakovic and Schlick [11] used ESR (electron spin resonance) spectroscopy to study the action of this reagent on Nafion. This method allowed monitoring the disappearance of the ESR signal from Ti^{3+} and the appearance of other radicals during the reaction. The initial radical formed was HOO*. The disappearance of this radical above $-53°C$ was accompanied by the appearance of a mixture of TiOO* and peroxide radicals. The peroxide radicals were observed only with dry Nafion and were stable up to 370 K. A broad signal, believed to represent fluoroalkyl radicals formed by chain scission appeared after 14 days and increased in strength after 92 days.

It may be mentioned incidentally, that another analytical use of 30% hydrogen peroxide is for the detection of traces of high surface area platinum: The catalytic effect of platinum causes rapid gas evolution.

Another test used hydrogen peroxide alone, particularly as a vapor at 120°C (Fig. 6.8). The authors of this poster conclude that the attack on Flemion® under these conditions occurs by an unzipping, starting at unstable end groups and by chain scission of the backbone. It should be noted that for a given amount of fluoride ion release, attack at unstable end groups (as well as attack on the branches) would result only in a very minor reduction of molecular weight, while backbone scission would result in a drastic loss of molecular weight even for a small release of fluoride ions. A further conclusion is that fluoride ion release is accelerated by ferrous ions, but occurs even in the absence of ferrous ions (Fig. 6.9). This could be restated

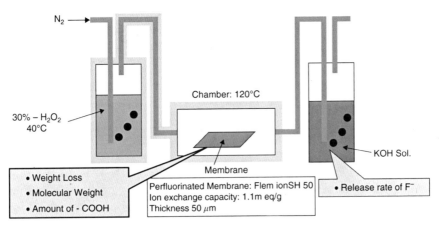

Figure 6.8 Membrane degradation using hydrogen peroxide vapors [22].

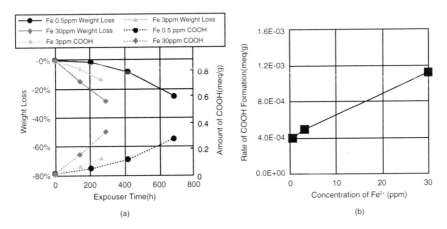

Figure 6.9 (a) Weight loss and amount of –COOH, and (b) rate of –COOH formation as a function of iron contamination [22].

as Fenton's reagent being more effective than hydrogen peroxide alone. In either case, it indicates the importance of avoiding iron contamination of fluorinated ionomers used in fuel cells, and the observation, that end of life MEAs are often contaminated with iron, nickel and other cations, is of some concern. Some degradation products were trapped inside the polymer and could be leached out by hot water. The main product among these was tentatively identified as the carboxylic/sulfonic acid derived from cleavage of the entire branch, either by attack on the oxygen linking the branch to the backbone or by unzipping the backbone past a branch point.

Minoru Inaba et al. in a recent publication (*J. Pow Sourc*, in print) used FTIR and NMR to examine Nafion 117 membranes after exposure to Fenton's reagent. They concluded that attack on the backbone and on the branches occurs at similar rates.

Damage to the MEA may occur not only as a result of chemical attack during operation, but also as a result of mechanical stresses induced as a result of humidity cycling, with or without electrical load. The ability to withstand humidity cycling is slightly improved by a reinforcement with Goretex®, but, at least for solution cast film, very much dependent on the casting conditions (Fig. 6.10).

6.4   Direct Methanol Fuel Cells (DMFCs)

One principle problem with hydrogen as a fuel is the difficulty of transporting, distributing and storing a gas that cannot be liquified at ambient temperatures. The use of methanol as a fuel has therefore attracted considerable interest. Perfluorinated ionomers, such as Nafion, exhibit high permeation rates for methanol and other polar solvents. This methanol crossover is increased by the flow of current. It represents not only a loss of fuel, but also to the extent that the oxidation of permeated methanol at the cathode competes with the cathodic reduction of oxygen, it diminishes the available potential at the cathode and therefore the cell voltage.

Recently, there was a interest in the use of layers of nanoparticles to reduce methanol crossover:

Tang et al. [12] minimized the methanol crossover by a self-assembled coating of a layer of palladium nanoparticles on the surface of Nafion. A loading of 0.002 mg/cm^2 Pd reduced the methanol crossover from 1.4×10^{-7} to 1.3×10^{-15} mol/cm$^2 \times$ s. There was no reduction in the proton conductivity of the membrane. A similar approach was taken by Kim et al. [13], except that the palladium layer was created by a chemical reduction method. Sun et al. [14] created a palladium layer on Nafion by electroless

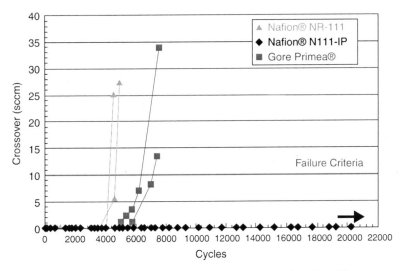

Figure 6.10 Ability of solution cast Nafion products to endure humidity cycling (courtesy Tom Greszler, Paper presented at the High Temperature Working Group Meeting, San Francisco, September 2006). Homogeneous membranes: Dupont™ NR-111 –25 μm, 1100EW Nafion®; Ion Power™ N111-IP –25 μm, 1100EW Nafion®. Composite membranes: Gore™ Primea® Series 57 (expanded PTFE filled reinforcement). Humidity cycling accelerates mechanical failures in the absence of electrochemical degradation. Different processing methods for the same polymer dramatically effects humidity cycling durability. Mechanical reinforcement insufficient to prevent humidity cycling induces crossover leak.

plating, and Ma et al. used sputtering to deposit a 1 nm layer of palladium alloys (or about 0.0012 mg/cm^2) on Nafion [15]. Mu et al. self-assembled a layer of 4 nm gold nanoparticle on the surface of a Nafion film and reduced methanol crossover current from 168 to 18 mA/cm^2 using 2 M methanol at 60°C [16]. Liu et al. deposited nanoparticles of titanium dioxide on Nafion at a coating density of 0.009 mg/cm^2 and observed an improvement in cell voltage and power density [17].

Bae et al. [18] grafted polystyrene onto NAFION onto porous polypropylene and observed a reduction in methanol crossover.

Composites of Nafion and inorganic oxides have also received attention as an approach to reduced methanol crossover (compare Section 3.10 "Composite Materials of Ionomers and Inorganic Oxides"). This subject has been reviewed by Li [19]. Xu et al. combined silica (from tetraethoxysilane) and phosphotungstic acid (from solution) inside Nafion [20].

The optimum silica content is 38 ppm. At 80°C, open circuit voltage was 750 mV and maximum power density 70 mW/cm^2 compared to 680 mV and 62 mW/cm^2 for commercial Nafion.

6.5 Manufacture of MEAs

As discussed, an MEA consists of three components: A central iono-mer film, typically 20–100 μm thick, coated on both sides with the anode and cathode electrode (or catalyst) layers, each about 10 μm thick. The two electrode layers may not be identical: the cathode layer frequently contains a higher platinum loading than the anode layer because of the more difficult cathode reaction.

The electrode layers are usually prepared by casting an "ink" on a support film or decal. The ink consists of a suspension of the catalyst (platinum on carbon black) in a liquid composition of the ionomer (the pigment and vehicle, respectively, in paint terms). A typical composition of the electrode layer after the removal of solvent is 50% ionomer, 25% platinum and 25% carbon black. In some cases, the electrode layer "mudcracks" on drying; the layer is then held together only by the decal.

Hot pressing, followed by removal of the two support films from the electrode layers then laminates the three components. In many cases, the central ionomer film is made larger than the electrode layers, creating a "frame" of ionomer film around the active area; or a "frame" of some nonconducting film, such as Kapton® polyimide, is introduced in the hot pressing step. The purpose of the frame is to limit the use of the expensive catalyst to the active area and to allow punching of holes for the gas distribution channels through the frame without the possibility of short circuiting the electrode layers.

A continuous process for making MEAs has been patented by Preischl et al. [21]. In a completely different approach to fuel cells, the catalyst is imbedded in the surface of a carbon fiber fabric serving as the gas-diffusion medium. The gas-diffusion medium then becomes the electrode. Such electrodes are commercially available from E-TEK (now a division of PEMEAS) under the name of "ELAT". The ELAT electrode incorporates a gradient of pore size and hydrophobicity from membrane to the gas (or back) side. More recently, a "fine gradient" (=fg) electrode, with reduced step changes in the gradient, has been introduced (Fig. 6.11).

A recent review article covers the various fluorinated ionomers used in fuel cells [24].

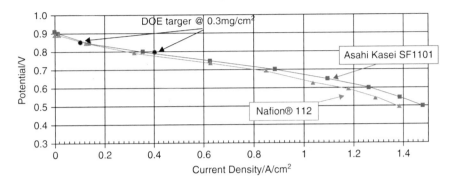

Figure 6.11 Performance of a fine gradient ELAT® electrode with 0.39 mg/cm total Pt. [23]. Alloy cathode, 0.28 mg/cm² 0.11 mg/cm² anode; cathode/anode: 250 kPa total, 80°C/85°C; Cell: 80°C.

References

1. Xie, Z., Navessin, T., Shi, K., Chow, R., Wang, Q., Song, D., Andreaus, B., Eikerling, M., Liu, Z., Holdcroft, S., J. Electr. Chem. Soc., **152**(6), A1171–A1179, 2005.
2. Srinivasan, S., Yang, C., Bocarsly, A., Tulyani, S., Benzinger, J., J. Membr. Sci., **237**(1–2), 145–161, July, 2004.
3. Lightner, V., 2005 Annual DOE Hydrogen Program Review, FC1.
4. More, K., 2005 Annual DOE Hydrogen Program Review, FC39.
5. Newman, J., 2005 Annual DOE Hydrogen Program Review, FC50.
6. Plug Power, Product Literature.
7. Borup, R., 2005 Annual DOE Hydrogen Program Review, FC40.
8. Hicks, M., 2005 Annual DOE Hydrogen Program Review, FC12.
9. Costamagna, P., Yang, C., Bocarsly, A., Srinivasan, S., Electrochim. Acta **47**(7), 1023–1033, 2002.
10. Otsuka, K., Hosokawa, K., Yamananka, I., Wada, Y., Morikawa, A., Electrochim. Acta, **34**(10), 1485–1488, October, 1989.
11. Bosnjakovic, A., Schlick, S., J. Phys. Chem. B, **108**(14), 4332–4337, 2004.
12. Tang, H.L., Pan, M., Jiang, S.P., Yuan, R.Z., Mater. Lett. **59**(28), 3766–3770, December, 2005.
13. Kim, Y.J.,Choi, W.C., Woo, S.I., Hong, W.H., Electrochim. Acta, **49**(19), 3227–3234, 2004.
14. Ma, Z.Q., Cheng, P., Zhao, T.S., J. Membr. Sci., **215**(1–2), 327–336, 2003.
15. Sun, H., Sun, G., Wang, S., Liu, J., Zhao, X., et al. J. Membr. Sci., **259**(1–2), 27–33, August, 2005.
16. Mu, S., Tang, H., Wan, Z., Pan, M., Yuan, R., El.chem. Commun., 7(11), 1143–1147, November, 2005.
17. Liu, Z., Guo, B., Huang, J., Hong, L., Han, M., Gan, L.M., J. Pow. Sourc., (in Press 2005).

18. Bae, B., Ha, H.Y., Kim, D., J. Membr. Sci., **202**(1–2), 245–252, June 2002.
19. Li, X., Roberts, E.P.L., Holmes, S.M., J. Pow. Sourc., **154**(1), 115–123, March, 2002
20. Xu, W., Lu, T., Liu, C., Xing, W., Electrochim. Acta, **50**(16–17), 3280–3285, May, 2005.
21. Preischl, C., Hedrich, P., Hahn, A., US patent 6,291,091 assigned to Ballard, September 18, 2001.
22. Hommura, S., 2005 Fuel Cell Seminar, Palm Springs, CA, Poster 103.
23. DeCastro, E., 2005 Fuel Cell Seminar, Palm Springs, CA, Poster 177.
24. Souzy, R., Ameduri, B., Prog. Polym. Sci., **30**(6), 644–687, June, 2005.

Further Reading

The Electrochemical Society has published a number of books on the subject of PEM fuel cells. These books contain papers presented at ECS meetings.

A) Proton Conducting Membrane Fuels Cells 4, ISBN 1-56677-434-9 (2004).
B) Proton Exchange Membrane Fuels Cells 5, ECS Transaction Vol. 1, No. 6, ISBN 1-56677-496-9 (2005).
C) Proton Exchange Membrane Fuels Cells 6, ECS Transaction Vol. 3, No. 1, ISBN 1-56677-501-9 (2006).
D) Durability and Reliability of Low Temperature Fuel Cell Systems, ECS Transaction Vol. 1, No. 8, ISBN 1-56677-491-8 (2005).

7 Commercial Membrane Types

In this chapter, some commercially available products will be described to help the reader in selecting a suitable membrane for a given application and in understanding its function. The manufacturers are reluctant to reveal the exact composition of their products, particularly of the more recently introduced ones. Nevertheless, a description of the older types will help the reader to understand the concepts involved. Web sites of manufacturers (such as www.Nafion.com) provide data on the available membranes and resins.

7.1 Unreinforced Perfluorinated Sulfonic Acid Films

These films are available in a variety of thicknesses and are typically made of either 1000 or 1100 EW polymer. In the Nafion® code, the first two digits indicate the EW and the remaining digit(s) the thickness in mils (1 mil = 25.4 µm).

The films listed in Table 7.1 are all made from extruded precursor film. More recently, DuPont has discontinued Nafion 112 and instead offers solution cast film of 25 and 50 µm thickness (NRE211 and NRE212, respectively; see also Appendix C).

7.2 Reinforced Perfluorinated Membranes

For most industrial electrolysis applications, fabric reinforcement is used to improve the mechanical strength, particularly the resistance to tear propagation. To match the chemical stability of the perfluorinated ionomer, pTFE is generally chosen. The details have been discussed in Section 3.4 "Fabrication".

7.2.1 Sulfonic Acid Membranes

Please refer to Table 7.2.

Nafion 324: A main layer of 125 µm, 1100 EW sulfonic acid polymer reinforced with a 10 × 10/cm leno weave of 400-denier brown multifilament yarn.

Table 7.1 Unreinforced Perfluorinated Sulfonic Acid Films

Type	Thickness (μm)	Equivalent weight (nominal)	Maximum width (m)
Nafion 105	130	1000	1.22
Nafion 112	50	1100	1.22
Nafion 1135	90	1100	1.22
Nafion 117	180	1100	1.22
Flemion FL-4	110	890	
Flemion FL-12	300	920	
Aciplex AC-4	110	980	
Aciplex AC-12	320	1080	

Table 7.2 Sulfonic Acid Membranes

Type	Fiber (denier)	Pattern	Thread count (per cm) warp × fill	Ionomer (μm)	Ionomer (EW)	Cation	Weight (g/m²)
N 324	400	Leno	11 × 10	130/25	1100/1500	H	480

This fabric has 32% open area (by IR transmission @ 1200/cm). On the cathode side of the membrane, there is a 25 μm barrier layer of 1500 EW sulfonic acid polymer for improved anion rejection. Recent production of this material appears to be calendered to reduce the overall thickness, although this information is not confirmed by the manufacturer. Nafion 324 is a general purpose industrial membrane particularly for applications requiring better anion rejection than Nafion 424.

Nafion 417: A single layer of 180 μm, 1100 EW sulfonic polymer reinforced with 16 × 16/cm plain weave of 400-denier brown multifilament yarn (essentially reinforced Nafion 117). This fabric has 25% open area. Nafion 417 has been replaced by Nafion 424, which uses the same leno weave as Nafion 324. Nafion 417 and 424 are general purpose membranes for applications where durability and strength are more important than anion rejection.

Nafion 350 and 450: These are now discontinued versions of Nafion 324 and 424, respectively. They used a white leno weave of a twisted ribbon of expanded pTFE as reinforcement.

7.2.2 Sulfonic/Carboxylic Membranes

These membranes are used exclusively for the electrolysis of sodium chloride. They represent by far the largest volume and value of all fluorinated ionomer products. The monomers used in their synthesis are shown in Table 7.3. All three manufacturers, DuPont, Asahi Glass and Asahi Kasei, use essentially the same constructions:

1) A 15–30-µm-thick layer of carboxylate polymer on the catholyte (cathodic electrolyte) surface of the membrane. The EW of this layer is usually about 1000, although it can be as high as 1200. This layer performs the critical function of OH ion rejection. Based on theoretical consideration, the minimum thickness for this layer is 10 µm.

2) An 80–130-µm-thick layer of sulfonate polymer of typically 1100 EW. The function of this layer is to anchor the carboxylate layer to the reinforcement and to protect it from the acidic anolyte.

3) A reinforcement layer made of pTFE filaments. The filaments used for this fabric are typically made by twisting a thin ribbon of expanded pTFE (GORE-TEX®) to a 200-denier fiber with 4–10 twists per cm. A true pTFE monofilament may also be used. The brown multifilament yarn made by the viscose rayon process was once used in Nafion 921, but is now obsolete for chlor-alkali membranes because of excessive electrical resistance (it is still used in Nafion 324 and 424). The fabric is either a plain weave with a thread count of about 10 threads per cm or a fabric of 3–6 threads per cm pTFE interwoven with 20–30 threads per cm of a sacrificial filament made of rayon or polyester. The sacrificial filaments are later destroyed to leave channels filled with anolyte, consequently increasing the conductivity of the membrane.

4) An optional coating of zirconium dioxide on the catholyte or on both surfaces of the membrane. This coating is applied as a suspension of zirconium dioxide in a solution of Nafion; it helps the release of gas bubbles from the surface of the membrane.

Table 7.3 Sulfonic/Carboxylic Membranes

Type	Fiber	Pattern	Count (per cm)	Sacrificial Fibers	Coated	Cation
N 90209	Tw.ribb	Plain	6 × 6	4 × 4	No	K
N 961	Tw.ribb	Plain	3.5 × 3.5	8 × 8	Cath.	K
N 954	Tw.ribb	Plain	6 × 6	4 × 4	Cath.	K
F 4221	Tw.ribb	Plain	8 × 8	None	Both	

Nafion 954: A main layer of 125 μm 1080 EW sulfonic polymer reinforced with a 6.5 × 6.5/cm plain weave of 200-denier twisted ribbon (4 twists per cm) of expanded pTFE, interwoven with 26 × 26/cm of a sacrificial fiber. A 38 μm barrier layer of 1050 EW carboxylic polymer plus a gas release coating on the cathode side.

NafioN 90209: This is same as Nafion 954 except but without the gas release coating.

Aciplex® F 4221 (Asahi Chemicals): (1) A main layer of 125 μm of 1030 EW sulfonic acid polymer reinforced with an 8 × 8/cm plain weave of 200-denier twisted ribbon (8 twists per cm) of expanded pTFE (no sacrificial fibers). (2) A 20 μm barrier layer of 1010 EW carboxylic polymer. A gas release coating is applied to both sides. It appears that this coating has been applied to the precursor film before lamination.

New membrane types are being developed for chlor-alkali membranes with the aim of incremental improvements in cell voltage and durability, and for fuel cell products with the aim of reduced cost.

8 Economic Aspects

8.1 Chlor-Alkali Cells

By far the largest application for fluorinated ionomers is as a separator in chlor-alkali cell. Figure 8.1 shows the worldwide chlorine capacity for the membrane process, as well as for the older technologies, for the years 1999, 2001 and 2003. There was a significant increase in capacity in the membrane process and a reduction in the other technologies. In 2003, the worldwide capacity by the membrane process was 20.3 million tons per year.

To convert this capacity into membrane terms, assume a current density of 5 kA/m^2 and a current efficiency of 95%. Under these conditions, 1 m^2 of active membrane area produces 151 kg of chlorine per day or 55 t per year. The active membrane area installed worldwide for chlorine production in 2003 was therefore about 370,000 m^2. The total area is slightly larger because of the inactive membrane area under the gaskets. At an average selling price (year 2005) of $850/m^2, the value of these membranes is about $320 million. With an average service life of slightly more than 3 years, the membrane replacement business is therefore about $100 million per year. An additional $20 million worth of membranes may be required each year for new capacity.

In terms of geographic distribution (Figs 8.2 and 8.3), chlorine production in the older industrial countries was either stagnant (Western Europe) or declined (North America), with a modest replacement of their older technologies (mercury and asbestos, respectively) by new membrane installations. Chlorine capacity in Asia increased substantially where membrane technology was the dominant process.

Figure 8.4 shows the market share of various technology sellers in new membrane chlor-alkali installations.

8.2 Fuel Cells

The current market for polymer electrolyte membrane (PEM) fuel cells can be divided into three main categories: (i) portable power, (ii) stationary power and (iii) transportation. This is further divided into five groups

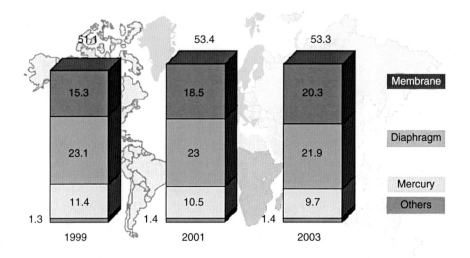

Figure 8.1 World chlorine capacity by process (millions of metric tons of chlorine) [1].

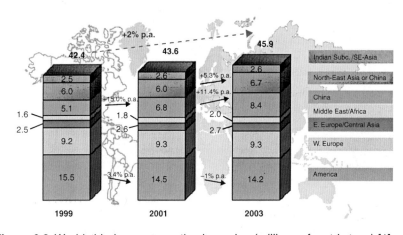

Figure 8.2 World chlorine consumption by region (millions of metric tons) [1].

as shown in Table 8.1. Portable and the related small stationary power together account for about 85% of all the installed units, transportation for about 12% and large stationary power for only a small fraction. For large stationary power, other fuel cell systems, such as phosphoric acid, or different technologies may be more economical.

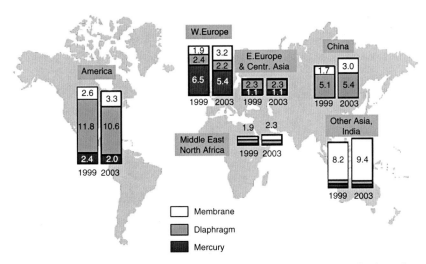

Figure 8.3 World chlorine capacity by type of process: regional distribution of type of processes, 1999/2003 (millions of metric tons of chlorine) [1].

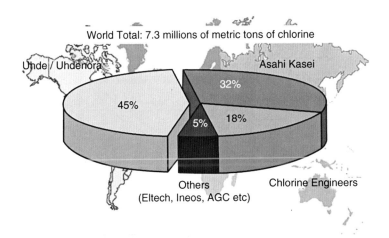

Figure 8.4 Market share of technology sellers [1].

Table 8.2 shows the number of companies engaged in different aspects of the development of PEM fuel cells.

There is hope that general automotive applications will be a very large market in the future, potentially larger than the chlor-alkali applications.

Table 8.3 lists the timetable of the large automobile manufacturers for fuel cell powered vehicles. The realization of these goals depends on the development of the necessary hydrogen infrastructure.

Table 8.1 Current Global Applications for PEM Fuel Cells (Percentage of Units Installed) [2]

Application	% installed
Portable power	67
Small stationary systems	18
Light vehicles	8
Buses and other transportation	4
Large stationary systems	3

Table 8.2 PEM Fuel Cell Activity by Function (Number of Companies in %) [2]

Activity	North America	Europe
Stack and system manufacture	25	25
Research and development	31	12
Component manufacture and supply	13	32
Testing/field trials	7	6
Other	24	25

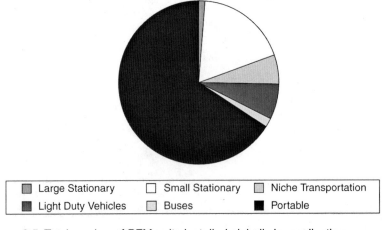

Large Stationary ☐ Small Stationary ☐ Niche Transportation
■ Light Duty Vehicles ☐ Buses ■ Portable

Figure 8.5 Total number of PFM units installed globally by application.

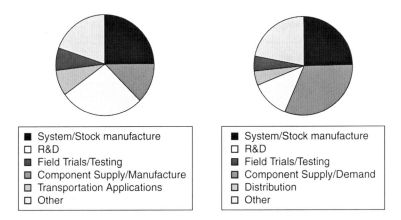

■ System/Stock manufacture	■ System/Stock manufacture
□ R&D	□ R&D
■ Field Trials/Testing	■ Field Trials/Testing
▨ Component Supply/Manufacture	▨ Component Supply/Demand
□ Transportation Applications	□ Distribution
□ Other	□ Other

Figure 8.6 PEM activity in North America (left) and Europe (right) by application; data based on a discrete sample of companies listed in the FCT Industry Directory.

Table 8.3 Timetable for Fuel Cell Powered Vehicles [2]

Manufacturer	Year	Number of Vehicles	Notes
DaimlerChrysler (Germany)	2012 2015	10,000	Initial launch, Mass Market
Ford (USA)	2015		"commercial readiness"
GM (USA)	2010–2015		Commercial viability
	2025		Mass Market
Honda (Japan)	2010 2020	12,000 (in USA) 50,000 (in USA)	Start production
Hyundai (Korea)	2010		Road tests 2009
Toyota (Japan)	2015		Will cost US$50,000

From: Crawley, G., Fuel Cell Today, March 2006.

References

1. 12th Krupp Uhde Chlorine Symposium, Dortmund, May 2004.
2. Crawley, G., Fuel Cell Today, March 2006.

9 Experimental Methods

9.1 Infrared Spectra

Infrared spectra are very useful for many purposes; for instance, equivalent weight (EW) determination, following chemical reactions of the functional groups, checking completeness of hydrolysis, etc.

The following bands have been observed for perfluorinated ionomers:

$-(CF_2)_n-$ (the polymer backbone): A weak band at 2360/cm, which is useful as a thickness band with which other bands are compared and a broad strong band from 1100 to 1300/cm.

$-SO_2F$ (the precursor): A weak band at 2710/cm useful for EW determinations and a very strong one at 1470/cm with a side peak at 1450/cm. This peak is useful for checking completeness of hydrolysis.

$-SO_2Cl$: A strong band at 1420/cm.

$-SO_3^-$ (the ionic form): A medium band at 1055/cm.

$-SO_2^-$ (the sulfinate salt): Bands at 940 and 1010/cm.

$-CO_2CH_3$ (the precursor): A strong band at 1780/cm. The free carboxylic acid has a similar absorption.

$-CO_2^-$ (the ionic form): A strong band at 1690/cm. The shift from 1780 to 1690/cm on neutralization, and back on acidification can be used to confirm a suspected carboxylic function.

$-COOH$: Carboxylate in precursor, acid or salt form, if it has more than one $-CF_2-$ group between the carboxylic function and the next ether linkage: a peak at 1020/cm. The absence of this peak can be used to recognize carboxylate polymers made by chemical conversion of the typical sulfonic precursor.

$-O-$ (the ether linkage): A strong band at 980/cm. Comparing the peak height and width of this peak with the sulfonate peak at 1055/cm may help to distinguish polymers with only one ether linkage (the Dow and 3M polymers) with those that have two ether linkages (compare the IR scans of the Nafion® and Dow polymers (Figs 9.2 and 9.3; also the precursor polymers Fig. 9.1a and b). In the ionic form, the two ether linkages can be resolved into one at 970/cm and another at 980/cm.

Water: A broad absorption between 3500 and 3700/cm and a peak at 1620/cm.

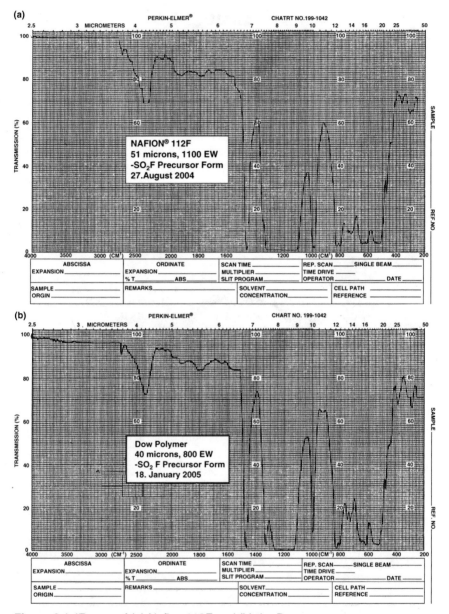

Figure 9.1 IR scan of (a) Nafion 112F and (b) the Dow precursor.

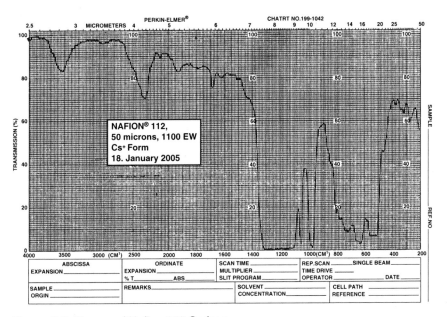

Figure 9.2 IR scan of Nafion 112 Cs form.

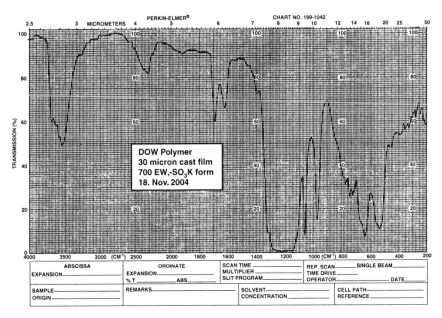

Figure 9.3 IR scan of Dow polymer, 700 EW, K form.

Ludvigsson has studied the IR spectrum of cast Nafion films (hydrogen form) at various temperatures and equilibrated at different relative humidity levels (Figs 9.4 and 9.5). He concluded that even with only one water molecule per sulfonic acid group complete transfer of the proton occurs [1].

Measurement of the transmission of a pTFE reinforcing fabric at a wavelength, where pTFE is opaque (8.5 μm or 1200/cm), can be used to determine the % open area of a fabric. The values obtained are slightly lower than the true values, as the light travels in a straight line whereas the electrical current can bend around fabric components, all these may not be in the same plane.

9.2 Hydrolysis, Surface Hydrolysis and Staining

Polymer precursors, sulfonyl fluoride and carboxylic ester, can be hydrolyzed using solutions of sodium or potassium hydroxide (NaOH and KOH, respectively). The solution can be aqueous or a mixture of an organic solvent, such as dimethyl sulfoxide (DMSO) and water to increase the rate of hydrolysis [2]. For applications where the introduction of sulfur compounds may be of concerns as a potential catalyst poison, lower aliphatic alcohols containing an alkoxy group can be used instead [3]. NaOH gives

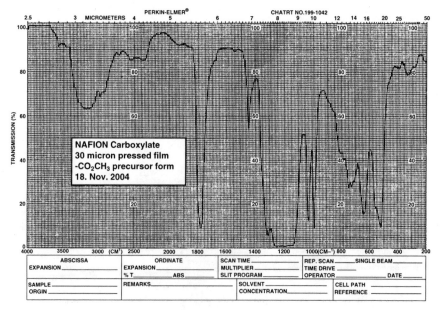

Figure 9.4 IR scan of Nafion carboxylate precursor, methyl ester form.

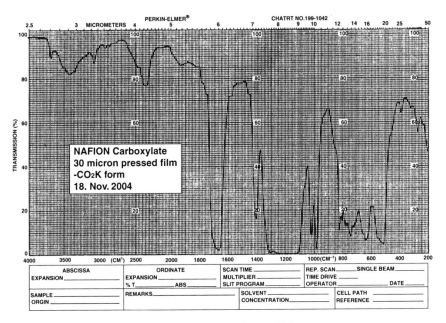

Figure 9.5 IR scan of Nafion carboxylate polymer, K form.

slightly higher reaction rates than KOH and is more easily exchanged against hydrogen ions. However, under adverse conditions (thick films, concentrated NaOH), use of NaOH may result in precipitation of byproduct NaF inside the polymer. While this material will leach out during the subsequent acid exchange, a defect will remain where NaF crystals are formed. KOH, on the other hand, has a broader range of solubility in mixed organic/water solvents and forms a more soluble fluoride salt.

The reaction proceeds with a sharply defined boundary. The progress of this boundary can be observed by slicing and staining a cross-section of film. If the hydrolysis is conducted at room temperature, a few sulfonyl fluoride groups, presumably encapsulated by the fluoropolymer component, will escape hydrolysis. At a point where slicing and staining indicate that a film has been totally hydrolyzed, IR analysis will indicate the presence of some sulfonyl fluoride groups by a peak at 1452/cm. Consequently, water absorption and conductivity will not have reached their final values. Hydrolysis at elevated temperatures is needed to complete the reaction.

Once hydrolyzed, there is no easy way back to the sulfonyl fluoride precursor. The carboxylic ester precursor, on the other hand, can be easily obtained by acid-catalyzed esterification of the hydrolyzed ionomer, or by reaction with trialkyl orthoformates [4].

Sometimes it is desirable to hydrolyze only one surface of a precursor film. This can be accomplished by using the precursor film as a liner in a dish and pouring the hydrolysis solution into this liner; or the film could be folded over itself and sealed around the edges. This bag could then be immersed in the hydrolysis solution or the solution could be injected into the bag.

Fluorinated cation exchange polymers can be stained by cationic dyes. This can be used to follow the depth of hydrolysis, to more clearly recognize the cross-section of a composite membrane, to distinguish between carboxylic and sulfonic layers, and to recognize the ionomer against a light-colored background, such as Teflon® or a catalyst support. In staining the cross-sections of films, it should be noted that slicing or microtoming must be done *before* staining. Attempts to recognize Nafion against a black background (carbon black) using a fluorescent dye in combination with UV light have had only limited success.

A wide choice of dyes are available including the following: Methylene blue stains Nafion almost black. Sevron® Brilliant Red 4G (DuPont) is often used to obtain a more transparent stain. Triphenyl methane dyes can be used to distinguish between carboxylic and sulfonic groups. In these dyes, two or all three of the phenyl groups are substituted in the *para* position by alkyl amino groups. One of these substituted phenyl groups is quinonoid with a positive charge on the nitrogen. Resonance permits this charge to be distributed among all the alkyl amino groups as well as the central carbon (carbonium ion). Examples include Malachite Green (two dimethyl amino groups), Methyl Violet (one additional monomethyl amino group) and Crystal Violet (three dimethyl amino groups). In an acidic solution, some or all of the benzoid amino groups become protonated, thereby preventing them from participating in the resonance. At a pH below 2, the sulfonic acid ionomer stains yellow, corresponding to the state where all amino groups carry a positive charge and only the central carbon is participating in the resonance. Carboxylic polymer will stain bluish green, corresponding to a state in which one benzoid amino group is not protonated and therefore able to participate in the resonance. (In the violet state, these dyes have two unprotonated benzoid amino groups.)

Bromthymol blue is particularly interesting. This indicator is blue at a pH above 7.5, yellow below pH 6 and exhibits various shades of green, presumably mixtures of yellow and blue, in between. When a sample of Nafion is immersed in the yellow solution of this indicator in 0.02–1 N HCl, the Nafion is stained deep burgundy red. The red stain of the Nafion sample further intensifies on drying. This is the color of the indicator in a concentrated, strong acid (>3 N HCl), a fact not generally recognized in the literature. The indicator therefore exhibits a different color in the polymer phase because of the higher acid concentration in this phase. Because this indicator does not stain the carboxylate ionomer, it can be used to distinguish the two polymers of a chlor-alkali membrane. If a piece of Nafion, stained red in this fashion, is immersed in 0.1 N NaOH, the change in color is first to yellow, then green and finally blue, progresses from the edges of the sample toward the inside. This has been viewed as an indication that the mass transport in extruded Nafion is faster in the plane of the film than in the thickness direction. A more likely explanation is that the mass transport in Nafion in all directions is faster than the transport through the solution/polymer boundary.

Dyes that do not stain the ionomer, when dissolved in a low surface tension solvent, can be used to make the fabric component of reinforced membranes more visible. Examples include Oil Red (Aldrich Chemicals) dissolved in kerosene, or potassium permanganate dissolved in 50% acetic acid with a small addition of a perfluorinated surfactant such as perfluorooctane sulfonic acid (FC 95, available from 3M).

If applied to the anolyte side of the membrane, these solutions will quickly wick into the sacrificial channels and into the void spaces within a multifilament yarn. Wicking takes place more slowly into the twisted Goretex® ribbons that are now commonly used as reinforcement. Since this test works only on the anolyte side of the membrane, it can also be used as a test to distinguish the two surfaces.

Inadvertent discoloration may occur if ionomers in the free sulfonic acid form are exposed to certain organic vapors. Acetone is particularly bad, but plasticizers used for polyvinyl chloride (PVC) and other plastics, solvents used in adhesives, organic vapors released from pine wood and some other woods can also cause discoloration. It is believed that these compounds will undergo acid-catalyzed condensation reactions within the polymer. As a result, if a roll of ionomer film in the free sulfonic acid form is exposed for months to laboratory or household air, then it will turn yellow or brown in the outer layer and in the exposed edges of the inner layers. It is, of course, not the ionomer itself that becomes discolored, but the degradation products formed from absorbed organic vapors under the action of the strong acid. For most applications, this discoloration is of no consequence. However, for some sensor applications, the accumulation of condensed organic compounds causes a slow drift of the calibration.

After hydrolysis, many commercial products are converted to the hydrogen ion form. A strong mineral acid, such as hydrochloric or nitric, in about 10% concentration is suitable for this purpose. Except for very thick shapes, the reaction reaches equilibrium in a few seconds at room temperature. It should be noted that only an equilibrium conversion is reached (see Section 9.4) and to reach near-total conversion, a second exchange using fresh acid is advisable. DuPont's specification for Nafion products in the acid form is a minimum 95% conversion to the acid form. After the acid exchange, the polymer is rinsed with distilled water until neutral and dried.

9.3 Other Reactions of the Precursor Polymer

Sulfonyl fluoride precursor form can undergo reactions with many reagents.

They include reactions with ammonia or amines, with anhydrous ammonia, either at dry ice temperatures, at the boiling point of ammonia or with ammonia in the gas form where a sulfon amide is formed [5]. As this material is a weak acid, the actual product is the ammonium salt of the sulfon amide $SO_2NH \cdot NH_4$. Substituted sulfon amides made from primary amines are even weaker acids [6]. Di- or tri- and polyamines, such as ethylene diamine, triethylene tetramine, etc., will cause some crosslinking after heating [7]. Reaction of the precursor polymer with 3-amino propyl triethoxy silane has been used to introduce crosslinking in Nafion [8]. The presence of water in any of these reactions will result in the formation of some ordinary sulfonate groups.

Reduction of the sulfonyl fluoride group to a sulfinic acid group can be accomplished with hydrazine.

The sulfinic acid can then be oxidatively cleaved with evolution of sulfur dioxide. The remaining fluorocarbon radical will convert to a carboxylic acid [9]. The oxidation is catalyzed by cupric and vanadyl ions:

$$-O-CF_2-CF_2-SO_2H + \tfrac{1}{2}O_2 + H_2O \rightarrow -O-CF_2-CO_2H + 2HF + SO_2$$

A similar conversion of the sulfonyl fluoride precursor to a carboxylic acid can be accomplished in a single step using lithium borohydride in tetrahydrofuran as a reducing agent. The sulfonyl chloride precursor can also be reduced with hydroiodic acid. In both the cases the sulfur is eliminated as hydrogen sulfide [10]. Another path to convert the sulfonyl fluoride precursor to a carboxylate polymer uses the reaction of the sulfonamide form with nitrous acid.

Creating a carboxylic barrier layer by a chemical surface treatment of a sulfonic base polymer would result in a "monolithic" structure, which is resistant to delamination. Unfortunately, a carboxylic functional group that is separated from an ether linkage by only a single $-CF_2-$ group, as the one shown in the equation, has poor long-term stability in hot concentrated caustic.

The sulfonyl fluoride precursor can be converted to an *anion* exchange polymer by reaction with a diamine containing both a secondary and a tertiary amine group, such as *N*-methyl piperazine. The presence of even trace amounts of water is particularly harmful in this reaction, because each sulfonate group formed will internally bind one anion exchange group. Another approach to fluorinated anion exchange polymers starts with the methyl ester precursor of the carboxylic ionomer. Reaction with a secondary amine, such as dimethyl amine, converts this to a substituted carboxylic amide. This is reduced to the tertiary amine, which can then be quaternized using methyl iodide [11].

9.4 Ion Exchange Equilibrium

In discussing ion exchange equilibrium, it is frequently necessary to express the concentration of ions in the polymer phase, requiring somewhat arbitrary assumptions of the volume of electrolyte occupying the polymer phase. This is illustrated by an example. An amount of 1046 g of Nafion sulfonic acid polymer with an EW of 1046, corresponding to a comonomer ratio of 6, may pick up 180 g of water (=10 mol per ion exchange site). This would be only partial hydration, for instance due to exposure to a fairly concentrated external solution or to partial conversion to a less hydrated cationic form (such as potassium). The total volume of the system would be about 670 ml and the concentration of sulfonic ion groups therefore 1.49 M. If this value is used for calculating the Donnan exclusion of anions, values much lower than those observed experimentally are obtained. Because of the phase separation between the ionic and the pTFE phase, one could argue that the pTFE phase represents a Teflon container holding the electrolyte and should therefore not be included in calculating the volume of this electrolyte. If the 600 g of pTFE plus the 81 g representing the trifluoro vinyl group are excluded, an electrolyte volume of 330 ml is obtained, resulting in a 3.03 M concentration of fixed ionic charges. The agreement with experimental values is then better, but still not satisfactory. Going a step further, one could also exclude the HFPO linkage next to the backbone (166 g or about 80 ml). This would be comparable to changing from a container made of Teflon homopolymer, to one made of Teflon PFA. The fixed ionic concentration would then be 4.0 M. Experimental values indicate a slightly higher concentration.

9.4.1 Anion/Cation Equilibrium (Donnan Exclusion)

The Donnan theory describes the distribution of anions and cations between the polymer phase and an external electrolyte. The partially hydrated polymer described above in equilibrium with a 0.1 M solution of NaOH is used as an example. It is assumed that the sodium and hydroxide ions are the only ions present (i.e., the polymer is in the sodium ion form). This equilibrium is dynamic: that means sodium ions cross the boundary between the two electrolytes back and forth due to Brownian motion.

Because of the higher concentration in the polymer phase, the initial rate of sodium ions leaving this phase is higher than the rate of those returning. This creates a negative potential on the polymer phase which retards the further loss of sodium ions from this phase and accelerates the return of these ions from the solution. Equilibrium is reached when the forces of the electrical potential balance the driving force due to Brownian motion and the rate of sodium ions leaving and returning becomes equal. The magnitude of the electrical potential is proportional to the logarithm of the ratio of sodium ion concentration outside and in $V = n \ln [Na]/[Na*]$. (The *inside* concentration is indicated by *. A larger inside concentration results in a negative potential of the polymer phase vs. the solution. The factor n equals RT/F for monovalent ions.)

This same electrical potential repels anions trying to enter the polymer phase. The concentration of OH ions inside the polymer is therefore lower than in the external solution. An analogous relationship between the potential and the concentration ratio exists: $V \sim n \ln [OH*]/[OH]$. (The negative potential of the polymer phase results in a smaller inside concentration of OH ions.) The two equations may be combined to:

$$[Na]/[Na*] = [OH*]/[OH] \text{ or } [Na] \cdot [OH] = [Na*] \cdot [OH*].$$

These are referred to as the Donnan equations.

For the above case of an ionomer containing a 4 M concentration of fixed ionic charges in equilibrium with a 0.1 N NaOH solution, the Donnan equation predicts $[OH*] = 0.0025$ and $[Na*] = 4.0025$, a very substantial exclusion of anions. For a more concentrated external solution, the exclusion would be poorer. However, as the concentration of the external solution increases, the ionomer becomes more dehydrated, resulting in a higher concentration of fixed ionic charges.

If a cation-selective membrane separates solutions of different compositions, the cations in the two solutions will readily exchange within the requirement of electrical neutrality. For instance, trace quantities of cations in a very dilute solution can be extracted and concentrated into a more concentrated acid solution. The effect is driven by the tendency of the hydrogen ions in the more concentrated solution to diffuse into the dilute solution. This creates a Donnan potential which forces the other cations in the opposite direction (i.e., Donnan dialysis).

9.4.2   Cation/Cation Equilibria

Ion exchange polymers exhibit preferences for one counter-ion over another. In fluorinated ionomers, such preferences are largely caused by differences in the hydration of various ions, the less hydrated ones being the more preferred. The equilibrium between a cation M and the hydrogen ion can be described in terms of a distribution factor K by the equation: $[M*]/[H*] = K [M]/[H]$. $[M*]$ denotes the concentration of cation M inside the polymer phase. Note that the factor K is not constant over the entire range of $[M]/[H]$ ratios because not all exchange sites have the same preference for M. The high preference sites are filled up first at low loadings of M. On the other hand, high loadings with a large cation will result in sharply reduced swelling of the resin with correspondingly higher selectivity. K values for some cations as a function of loading is shown in Fig. 9.6 [12, p. 28].

More extensive data exist for the distribution coefficients of conventional, polystyrene-based, cation exchange resins.

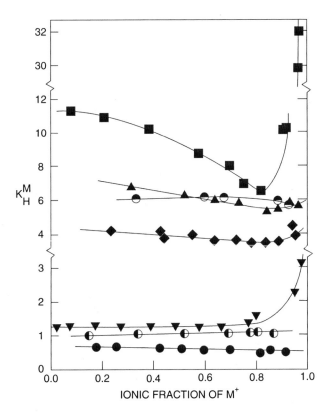

Figure 9.6 Distribution factor K for various cations vs. protons [12]. (As a function of ionic fraction of cation M, measured on 1200 EW Nafion at an ionic strength of 0.01 M.) ●, Li⁺; ▼, Na⁺; ◆, K⁺; ▲, Rb⁺; ■, Cs⁺; ◖, Ag⁺; ◗, Ti⁺.

Some of these data are shown in Appendix 4. While the data are for crosslinked sulfonated polystyrene, similarities to fluorinated ionomers exist because the dependence of the distribution coefficients on liquid phase, such as the formation of anionic complexes. The distribution of ferric ion in a hydrochloric acid solution compared to a perchloric acid solution is an example: In the hydrochloric acid solution, a pronounced minimum distribution coefficient is observed at 3 N HCl, with the distribution coefficient becoming less than 1 between 2 and 4 N. That means ferric ion is rejected from the resin because of the formation of an anionic chloride complex. Similar effects can be seen for Zn and Cd and some other metals.

Some of these selectivity differences are modest; however useful separations are achieved using a multistage counter-current electrodialysis process. This is illustrated in the following sketch; the top view of the cell stack is used to show the analogy to a distillation column (Fig. 9.7). Part of the catholyte product enriched in KOH is used as a "reflux" to sweep NaOH toward the anode compartment. The MP cell made by ElectroCell is quite suitable for this application because it is equipped with for manifolds. This number can be doubled by installing a gasket that will block all manifolds in the center of the stack. Eight different electrolytes can then be fed into the stack (four from each end).

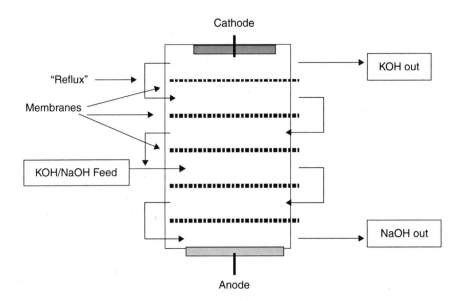

Figure 9.7 Multistage separation of cations (top view of the stack).

It should be noted that contrary to the exchange equilibria discussed earlier, the electrodialytic separation depends on the transport fraction of the different cations, which may be slightly different than the static distribution factors (K values).

K values for large organic cations, such as tetrabutyl ammonium or certain dyes, are quite large. It is almost impossible to remove them by ion exchange; the preferred method of removal is chemical destruction. The Hofmann degradation (boiling in a solution of NaOH) for instance will convert quaternary ammonium salts to tertiary amines, which are not ionized at the high pH and are therefore expelled from the resin. After careful rinsing, the polymer, now in the sodium form, can easily be converted to other ionic forms.

Fluorinated ionomers containing large organic cations, such as tetrabutyl ammonium, resemble to some extent the precursor polymer. They are soft and pliable and more hydrophobic. Their water content, and therefore their conductivities are low, and their melting point is significantly lower than that of other ionic forms.

9.5 Determination of EW by Titration or Infrared Analysis

Titration is used to calibrate other analytical methods. Its major limitation is the difficulty of drying the polymer in the proton form to a known water content value. Drying in a vacuum oven will leave at least one molecule of water (as a H_3O^+ ion) in the polymer. If the sample is in the form of a single piece of film, it is best to determine its weight after titration, when the polymer is in the salt form. A neutral salt, such as potassium sulfate, is added to speed up the titration. Potassium sulfate is preferred as the neutral salt over sodium chloride, because the potassium form of the ionomer is more readily dried than the sodium form, and the sulfate ion removes hydrogen ions from the exchange equilibrium by conversion to bisulfate, thereby speeding up the titration. The following may serve as an example:

About 1 g of sample, 50 ml of water and 300 mg potassium sulfate are placed in a beaker on a magnetic stirrer/hot plate. Bromthymol blue is added as an indicator. The mixture is titrated using 0.1 N NaOH at 50–70°C until a blue end point is reached. The sample is removed, rinsed with water, blotted, dried in a vacuum oven at 110°C and weighed. (Calculation: EW = mg sample × 10/ml NaOH. This gives the EW in the K form.) The EW in the hydrogen form is 38 units less.

A number of titrations of samples of different EWs can form the basis of a calibration curve (peak height ratio 2710/2360 vs. EW) for the faster infra red analysis of the precursor resin.

9.6   Determining Melt Flow

Next to EW of an ionomer the melt flow (MF) of precursor polymer is an important characteristic of an ionomer. It can be conveniently be determined by a device called a "melt indexer" (Fig. 9.8).

It consists of a heated vertical barrel in which a sample of molten polymer is forced out of an orifice by a piston loaded with a known weight. The extrudate is cut off from the orifice at time intervals (typically 10 min) and weighed. The relationship between the dimensions of the melt indexer, the viscosity of the polymer, the force applied and volumetric flow rate for a Newtonian liquid is given by the Hagen–Poiseuille equation [13]:

$$\dot{V} = \frac{\pi R^4}{8L}\frac{1}{\eta}\Delta P$$

Typical dimensions for the melt indexer are: piston diameter = 9.5 mm, orifice diameter (R) = 2.095 mm, orifice length (L) = 8.0 mm and the force applied to the piston is 1200 g. The temperature is of course the most important variable and needs to be maintained within a fraction of a degree.

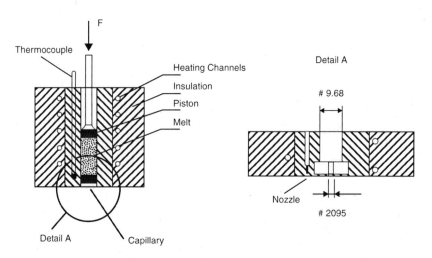

Figure 9.8 The melt indexer [13, p. 183].

Enough time should be allowed to let the device reach temperature equilibrium. Typical temperatures for the Nafion precursor are 270°C for 1100 EW, 290°C for 1500 EW and 250°C for 1000 EW and lower.

Most precursor polymers do not exhibit Newtonian flow but exhibit lower viscosities at higher shear rates (*shear thinning*). This effect is more pronounced in polymers of a broader molecular weight (MW) distribution. Measuring MF at different piston forces will therefore give some indication of the MW distribution. Another indication of broad MW distribution is the amount of swelling observed as the molten polymer exits the dye. Higher MW components of a polymer which has a broad MW distribution cause a degree of elastic deformation by the forces exerted on the melt during extrusion. As the melt emerges from the die, the stored elastic forces cause the extrudate to swell.

The MF of precursor polymer is usually determined on samples taken immediately after polymerization to make adjustments to the polymerization process. MF is again determined after pelletizing to select polymer lots of an MF suitable for extrusion. Blending of lots that fall above and below specification can be used to bring, to some extent, the MF into the specified range.

The information about melt viscosity is not only important for selecting polymer lots suitable for extrusion, but also as a predictor of the properties of the finished ionomer. In general, a low MF, that means a high MW, is desirable for the best end use properties. Low MF of the precursor form will result in good mechanical properties of the finished ionomer.

9.7   Distinguishing the Precursor Polymer from Various Ionic Forms

Sometimes the ionic state of a polymer sample is not clear. For samples in film form there are a number of easy tests:

1) *Precursor form*: A felt point marker will write on hydrolyzed polymer normally. If the sample is in the proton state some markers will change color in about ½ s. On precursor form the ink will form small beads. Some water placed on the surface of precursor polymer will bead up; and hydrolyzed polymer will be wetted. Small drops of water will form "goose bumps" and larger areas will curl. Infrared analysis will detect even traces of precursor polymer; it is best to neutralize any free acid groups using potassium

bicarbonate to obtain a better scan. Is this batch of pellets completely hydrolyzed? Place a sufficiently large sample in a 1 : 1 mixture of water/isopropanol to cover the pellets completely as they expand. Heat to just below boiling point. The hydrolyzed pellets will expand about ten times in volume and become almost invisible because of the match in refractive index. A single unhydrolyzed or partially hydrolyzed pellet in a batch of more than 10,000 pellets can be easily detected.

2) *Hydrogen or some other ionic form*: Place a drop of water on a film, pellet or powder sample. Put a pH paper in the drop. It should test neutral unless the sample was incompletely washed after an acid exchange. Now put some neutral salt (NaCl or KCl) in the drop of water, wait a few minutes and test the pH again. The pH will be < 2 even if only part of the exchange sites is in the hydrogen form. To detect more substantial fractions of hydrogen form, the sample is immersed in a solution of sodium or potassium bicarbonate. A hydrogen ion content of the sample is indicated by carbon dioxide evolution. If the EW of the sample is known, titration will give a quantitative answer to the question of what fraction of the ionic sites has counter-ion of hydrogen. Or the other cations are exchanged against a small volume of 10% HCl. Evaporation of the liquid to dryness will yield the foreign cations as chlorides, which can be further analyzed or weighed.

9.8   Fenton's Test for Oxidative Stability

The Fenton reaction was originally used as a preparative organic synthesis tool for the oxidation of α-hydroxy acids to the corresponding α-keto acids or in the oxidation of 1,2-glycols to hydroxy aldehydes [14,15]. It involves the use of a redox couple consisting of hydrogen peroxide and ferrous ion. As applied to fluorinated ionomers, it detects oxidative instability of the polymer through the release of fluoride ions.

There are several modifications of this test. One procedure has been published by Curtin *et al.* [16]:

The sample is treated with a solution of 30% hydrogen peroxide containing 20 ppm ferrous ions at 85°C for 16–20 h. The fluoride ion content of the solution is then determined using a fluoride-specific ion electrode. The same sample is then treated two additional times using fresh peroxide/ferrous solution. The results are expressed as total milligram of fluoride per gram of sample.

In another procedure, the samples are first converted to the ferrous ion form by immersion in a solution of ferrous sulfate. By locking the ferrous ion inside the polymer, the ineffective destruction of hydrogen peroxide in the liquid phase is minimized. Several samples can be treated in the same container as long as they can be removed and recognized separately. The samples are removed from the solution and heated separately in a 5% solution of hydrogen peroxide to 80°C for 6 h. The solution is then analyzed for fluoride ions. In the case of Nafion, only traces of fluoride ions are observed even under repeated tests. Also compare Section 6.3.

It is interesting to note that perfluorinated ionomers have also been used in the Fenton reaction in a completely different manner: as a chemically stable matrix to hold ferrous ions in the catalyzed oxidation of hydrocarbons either by hydrogen peroxide or by photo-chemical oxidation. These reactions are discussed under Section 5.7.

9.9   Examination of a Membrane

Occasionally it is necessary to examine a used membrane or an unknown new one:

(a) *Reinforced membrane*: Visual inspection will identity the reinforcement by weave pattern and thread count. It will also indicate the presence of scale and other contaminants. In the absence of a gas-release coating it will usually allow the determination of anode and cathode sides. The cathode side is glossy but shows the surface undulation of the reinforcement. The anode side is flat and dull and in many cases shows the imprint of the fibers of the release paper. A soft pencil will leave a faint image on the anode side, but not on the cathode side; on the cathode side the surface undulations can be felt in an attempt to write. In difficult cases, a solution of a dye in a low surface tension hydrocarbon solvent may help to identify the anode side (see Section 9.2). If the membrane has a carboxylic barrier layer, selective staining of a cross-section will identify this layer and therefore the cathode side (see Section 9.2). Or the differential swelling on changing from the salt to the free acid form will reveal the carboxylic (or other weak acid) layer. For this it is important to first immerse the test strip in a very dilute (<0.1 N) solution of NaOH. This may cause the strip to curl in either direction because the reinforcement may not be exactly centered. The NaOH solution is then replaced with very dilute acid. The change of curvature, as the carboxylic layer converts to the non-ionized and therefore not swollen state, is observed. Addition of ethanol will enhance the effect.

The machine (= MD) and transverse (= TD) direction can often be determined by examining the fabric: in a leno weave, the twisted double strand is always in the MD. Also, the MD yarns (the warp) are always straight and parallel, while the TD yarns (the fill) are sometimes curved or at some angle other than 90° relative to the MD. While any visible line will usually run MD, some lots of Nafion 324 and 350 exhibit "TD lines". These lines are caused by a change in surface undulation on the cathode side and are visible in reflected light. They are spaced 25 mm apart and run perfectly straight and parallel. At 45° between crossed polarizers they show up as dark bands (see Section 4.5).

Gas-release coatings applied as an "ink" can be removed by gently brushing with methanol. Some of the above-mentioned observations are easier after removal of the gas-release coating. Boiling of a multilayer membrane, that has been converted to the hydrogen form, in methanol or in a 1 : 1 mixture of isopropanol and water will delaminate the components. The barrier layer (carboxylic or high EW sulfonic) can be peeled off and the fibers of the reinforcement removed.

(b) *Unreinforced films*: The MD can usually be recognized by some fine lines visible in reflected light. If two sample strips are cut at right angles and gently stretched while placed at 45° between crossed polarizers, the one cut in the MD will lighten, while the one cut in the TD will first darken and then turn lighter (see Section 4.5).

9.10   Determining the Permselectivity

Permselectivity is a measure of the ability of a membrane to discriminate between anions and cations. It can be determined by measuring the concentration potential developed between solutions of the same electrolyte at different concentrations separated by the test sample (see Fig. 9.9). Potassium chloride is usually chosen as the electrolyte because the potassium and the chloride ion have equal mobility (see Table 5.1). An electrolyte with unequal ion mobility, such as sodium chloride, would develop a concentration potential even when separated by a non-selective separator, like a fritted glass.

The potential developed by an ideal selective separator is described by an equation similar to the Donnan equation: $P = RT/F * \ln C1/C2$, where R is the gas coustant, T is the absolute temperature, F is Faraday's constant, $C1$ and $C2$ are the concentrations of the two electrolytes.

For a ratio of concentrations of 10 : 1, this theoretical potential is about 58 mV. The permselectivity is then the measured potential expressed as a percentage of the theoretical potential of a 100% selective separator.

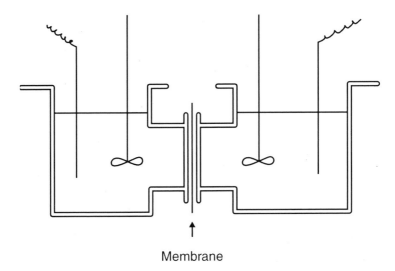

Membrane

Figure 9.9 Permselectivity cell [17].

It is important for the measurement that the two electrodes are identical. This can be verified by measuring the potential between the two electrodes when they are immersed in the same electrolyte. Or one can switch the electrodes during the measurement and take the average value of the potentials.

The permselectivity is independent of membrane thickness but is a function of the electrolyte concentration, as predicted by the Donnan theory. Some permselectivities for Nafion are:

0.1 N vs. 0.5 N KCl: 95%
0.5 N vs. 1 N KCl: 90%
1 N vs. 3 N KCl: 68%

9.11 Measuring Pervaporation Rates

A convenient method to measure pervaporation rates uses the Thwing Albert cup (Fig. 9.10):

The cup is filled with water or other liquid to be measured, weighed and placed in an inverted position on a suitable support that allows the exposure of the sample to a high velocity air stream. Weights are taken at periodic time intervals to determine the rate of pervaporation. If a liquid

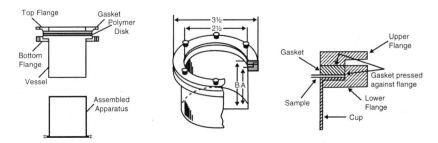

Figure 9.10 Thwing Albert pervaporation cup [13].

mixture, such as a water/alcohol mixture, is used, analysis of the residual liquid will give some indication of selectivity.

The main limitation of the method is in measuring very high rates of pervaporation, as they are typically encountered with polar liquids permeating through ionomer films. One should realize that even in the absence of a film sample to be tested, the rate of evaporation from the surface is limited and very much a function of the air velocity over the surface. To recognize this limitation, measurements should be taken at several different air velocities, and if possible, also at two different sample thicknesses. Another manifestation of high rates of pervaporation is the evaporative cooling and the resulting difficulty of maintaining a constant temperature. Still another manifestation is the stretching of the test sample as a result of the large loss of liquid volume. The stretched sample is of course thinner and has a larger surface area.

9.12   Simple Electrolytic Cells

For a batch reaction, a very simple cell can be assembled from a beaker placed on a magnetic stirrer/hot plate. A membrane bag placed inside the beaker would typically surround the "counter"-electrode (i.e., the electrode which is not of primary interest). A cylindrical "working" electrode is placed outside of the bag. In Fig. 9.11, the working electrode is an anode grid made of a lead alloy. The anode surrounds the dark membrane bag, which is supported by a plastic mesh. The end of the cathode is visible above the other components. This cell is intended for the regeneration of chromic acid. Periodic sampling of the "working" electrolyte in the beaker and the "counter"-electrolyte in the bag is used to follow the progress of the reaction.

Figure 9.11 Simple electrolytic cell.

For chlor-alkali research, a small cylindrical cell made of a glass anode compartment and a cathode compartment machined from polymethyl methacrylate has been very useful (Fig. 9.12). The design minimizes the catholyte volume. This is important because the current efficiency is determined by titration of the catholyte overflow collected during the passage of a certain number of ampere hours through the cell; any change in the inventory of NaOH retained within the cell would cause an error. The anode compartment in comparison is fairly large to maintain constant cell temperature (a well for a cartridge heater in the bottom of the anode compartment is not shown in the figure).

Saturated brine is fed into the anode compartment through the hollow anode stem (a piece of 6 mm titanium tubing); depleted brine overflows and is discarded. Anolyte concentration is determined by the rate of brine feed and can be measured by determining the density of the overflow. Water is fed through the hollow cathode stem; the overflowing catholyte is collected for titration. The result of the titration is the basis for calculating the catholyte concentration, the current efficiency and the water transport

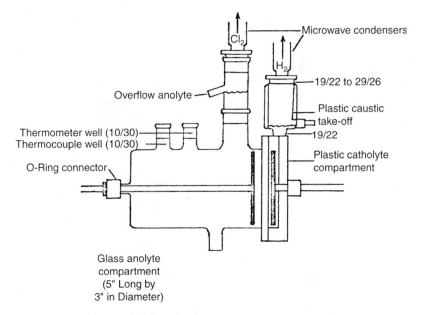

Figure 9.12 Laboratory chlor-akali cell [18].

per sodium ion passed, based on the ampere hours passed and the water feed during the sampling period.

The electrode–membrane spacing can be adjusted by sliding the electrode stems in and out through the threaded end fittings. The special components needed for the anode compartment (the 80-mm flange which forms the anolyte compartment, the clamp to hold the two compartments together and the threaded glass fitting for the anode stem) are available from Lab-Glass, Inc. A picture of the cell components is shown in Appendix B.

A small plate and frame-type electrolytic cell is available from Electro-Cell AB (the "MP cell"; see Table 5.2 and Fig. 5.7).

References

1. Ludvigsson, M., Lindgren, J., Tegenfeldt, J., Electrochim. Acta, **45**(14): 2267–2271, 2000.
2. Miyazaki, M., Hiyoshi, T., US Patent 5,066,682 assigned to Asahi Kasei, Nov. 19, 1991.
3. Banerjee, S., Grot, W., US Patent 5,310,765 assigned to DuPont, May 1994.
4. Grot, W., US Patent 4,415,679 assigned to DuPont, Nov. 15, 1983.
5. Grot, W., Resnick, P., US Patent 4,113,585 assigned to DuPont, Sep. 12, 1978.

6. Grot, W., US Patent 3,784,399 assigned to DuPont, Jan. 8, 1974.
7. Grot, W., US Patent 3,969,285 assigned to DuPont, Jul. 13, 1976.
8. Mauritz, K., Moore, B., et al., Polymer, **38**(6), 1345–1365, 1997.
9. Grot, W., Molnar, C., Resnick, P., US Patent 4,267,364 assigned to DuPont, May 12, 1981; US Patent 4,544,458 assigned to DuPont, Oct. 1, 1985.
10. Seko, N., *et al.*, US Patent 4,151,053 assigned to Asahi Kasei, Apr. 24, 1979.
11. Matsui *et al.*, Japanese Patents JP59-122520, 1984; JP60-84312, 1985; JP60-84313, 1985.
12. Eisenberg, A., Yeager, H. (eds) Perfluorinated ionomer membranes, ACS Symposium Series, 180, 1982.
13. Ebnesajjad, S., Fluoroplastics, Vol. 2: Melt-Processible Fluoroplastics (pp. 183, 369), Plastic Design Library, William Andrew Inc., Norwich, NY, 2002.
14. Fenton, H.J.H., Proc. Chem. Soc., **9**, 113, 1893; J. Chem. Soc., **65**, 899, 1894.
15. Walling, C., Amarnath, K., J. Am. Chem. Soc., **104**, 1185, 1982.
16. Curtin, D.E., Lousenberg, R., Henry, T.J., Tangeman, P., Tisack, M., J. Power Source., **131**, 41–48, 2004.
17. Grot, W., Munn, G., Walmsley, P., Paper presented at the 141th ECS Meeting, Houston, 1972.
18. Grot, W., Paper presented at the Congress of the GdCh-Section "Applied Electrochemistry", Darmstadt, Oct. 7, 1977. (Abstract in Chem-Ing. Tech., **50**(4), 299–301, 1978.)

10 Heat Sealing and Repair

While melt fabrication of fluorinated ionomers is typically done in the precursor form, the finished product in the hydrogen ion form has enough melt flow to permit heat sealing and repair. Impulse heating is used for heat sealing, because at the required temperatures the ionomers undergo slow thermal degradation. Ionomers in the tetra alkyl ammonium form have a much lower melting point than ionomers containing inorganic cations. The tetra butyl ammonium form is particularly easy to heat seal [1].

Commercial heat-sealing equipment is available from Vertrod and Quivers in Milam, Italy. Quivers offers a complete system for membrane repair, including equipment to check large sheets of membranes for leaks, automated patching equipment and patches made of Nafion® attached to a Kapton® release sheet.

Figures 10.1 through 10.3 show a defect in a large sheet of membrane being repaired. Most of the sheet is rolled up under the support arm of the heat sealer. Five precut repair patches can be seen in the foreground.

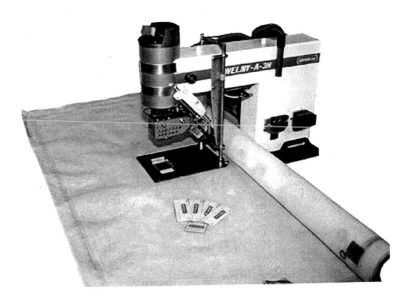

Figure 10.1 Quiver heat sealer applying a patch to a membrane (courtesy: Quiver Ltd., Milan, Italy).

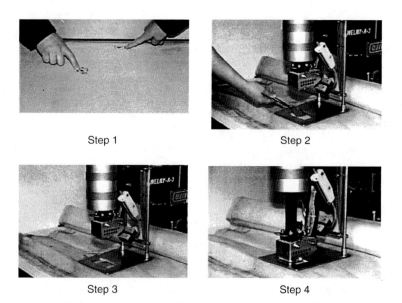

Figure 10.2 The first steps of the repair procedure (courtesy: Quiver Ltd., Milan, Italy). Step 1: The spots to be repaired are identified and marked. Step 2: The patch is placed inside the centering plate. Step 3: The operator pushes the two start buttons. Step 4: The welding cycle starts.

Figure 10.3 The final steps of the repair procedure (courtesy: Quiver Ltd., Milan, Italy). Step 5: Cooling starts automatically at the end of the heating cycle. Step 6: At the end of the cooling cycle the cylinder releases. Step 7: The Hosmy leak tester is placed over the patch to check the repair.
Step 8: If there is no leak, the centering plate is removed in preparation for the next repair.

Conditions suggested by Quiver are as follows:

Heating block temperature: 250–350°C
Heating cycle: 2–3 min
Cooling cycle: 1–2 min
Clamping force: 260–300 klb (or kg force)
Clamping pressure: 11–19 bar

For the repair of membranes with a cathode side barrier layer, it is important that the anode side of the patch is sealed against the anode side of the membrane to be repaired.

Reference

1. Grot, W., US Patent 3,884,885, assigned to DuPont, May 20, 1975.

11 Handling and Storage

Perfluorinated sulfonic acids (PFSAs) have unique properties and require special handling. Even though PFSAs are as thermally and chemically stable as polytetrafluoroethylene (PTFE), they are vastly different from PTFE. PFSAs are highly hydrophilic in comparison with total hydrophobicity of PTFE. PFSAs are acidic whereas PTFE is neutral. Skin contact with PFSAs is safe because of the immobility of acid groups inside the polymer structure. This chapter describes safe handling procedures and precautions that should be followed with perfluorinated sulfonic acids.

11.1 Handling the Film

Fluorinated ionomer products in the precursor form, in general, do not require special precautions in handling and storage. Products in the ionic form (hydrogen, sodium, potassium, etc.) are subject to mechanical damage, particularly when folded or creased. In addition, products in the free sulfonic acid form also tend to pick up contaminants from the environment. For more information about standard safety practices please refer to *Nafion® Safety in Handling and Use Bulletin* and the material safety data sheet. Chlor-alkali customers should refer to *Nafion® User's Guide*. All three guides are available from DuPont Company (www.Nafion.com).

11.1.1 Mechanical Damage

Damage by sharp objects obviously must be avoided. Sharp folding will leave a permanent set in a sheet. Most damaging is folding in two different directions (like a table cloth). For storage, large sheets can be rolled up on a core of at least 5 cm, preferably 10 cm diameter.

11.1.2 Chemical Contamination

Products in the free sulfonic acid form will pick up foreign cations on contact with tap water or other ion-containing solutions. In addition, they will pick up contaminants in vapor form, obviously any volatile base, such as ammonia or amines. Even then organic compounds that can undergo acid-catalyzed condensation reactions can cause discoloration of sulfonic acid products. Such organic vapors can be found in laboratory air (acetone) or can be released from packing materials (terpenes from wood, plasticizers from plastics, and so on). Solvents in the adhesive of labels can diffuse through a polyethylene bag and stain the ionomer underneath.

11.1.3 Cutting

A razor blade, knife or scissors may be used for cutting. Because membranes are usually installed wet, a decision must be made whether to cut wet or dry. Cutting dry is more convenient but requires careful estimation of the dimensional changes on wetting, particularly with respect to positioning bolt holes, etc. Unreinforced films exhibit significant dimensional changes as a result of even modest changes in relative humidity.

11.2 Pretreatment

Membranes change dimension based on moisture and cation (H^+ vs. Na^+) content. Installation of a dry membrane in an electrolyzer will, therefore, result in wrinkling after exposure to the electrolyte. It is the purpose of the pretreatment to obtain an operating membrane that is flat without wrinkles or excessive tension. For a better defined pretreatment condition, one must: (1) keep a record of the pretreatment conditions and the resulting expansion. (2) After use, inspect the membrane for wrinkles or excessive tension. (3) Adjust the treatment conditions (temperature, time) accordingly and compare the resulting expansion with the one originally recorded.

Chlor-alkali membranes sold in the potassium form should be pretreated with dilute NaOH or sodium bicarbonate solution as specified in the *Nafion®* *User's Guide* available from DuPont Company (www.Nafion.com).

Pre-expanded chlor-alkali membranes require no pretreatment. Membranes in acid form should be pretreated by soaking in hot water for at least 30 min and allowed to cool under water.

11.3 Installation

Install wet membranes promptly after pretreatment to prevent drying. Use water spray if needed during installation. Once clamped into the electrolyzer, membranes must not be allowed to dry out. Membranes pre-expanded with di-ethylene glycol must be rinsed with water prior to energizing to prevent foaming.

12 Toxicology, Safety and Disposal

12.1 Toxicology

The acute oral toxicity of perfluorinated ionomers is quite low. For rats the oral LD50 is higher than 20,000 mg/kg body weight. However, perfluorinated ionomers will release toxic gases when exposed to temperatures in excess of about 250°C. There are three likely scenarios for such an exposure:

1) Fire. While perfluorinated ionomers will not sustain combustion in air, they could be involved in a fire of other materials. Self-contained breathing apparatus should be worn under these conditions.
2) Heat sealing and repair exposes the polymer briefly to temperatures of about 300°C. Adequate ventilation should be provided for these operations.
3) Contamination of tobacco products.

12.2 Safety

A great deal of fluorinated ionomers have been safely used over the last decades. No injury has been reported during this time arising from handling or exposure to these materials.

12.2.1 Skin Contact

Testing perfluorinated sulfonic acids (PFSAs) in rabbits has not revealed irritation of the skin. Human volunteers underwent tests to determine skin irritation and sensitization potential of PFSA. The results revealed that no unusual dermatitis hazard in the normal use of membranes in industrial applications. Skin irritation to some individuals after prolonged contact may occur [1].

12.2.2 Thermal Stability

Fumes generated from the decomposition of plastics including fluoropolymers, such as polytetrafluoroethylene and perfluorosulfonic acid, bear health hazards to different extents. Fluoropolymers are generally more resistant to decomposition as a result they tolerate exposure to high temperatures than other thermoplastics.

Perfluorosulfonic acid copolymer with tetrafluoroethylene has a maximum operating temperature of 175°C under anhydrous conditions. In the presence of water and organic proton donating solvents, several days of resistance at temperatures in the range of 220–240°C has been reported.

12.2.3   Thermal Degradation Products

An infrared technique has been developed by DuPont for the analysis of thermal effluents [1]. This method was used to analyze the effluents of PFSA that evolve during overheating this polymer. A sample of 0.5 g polymer was heated in a steel tube where an airflow of 13 ml/min was maintained. The sample was heated at 10°C/min to 200°C, and then at 5°C/min to 400°C. It was held for 20 min at 400°C. Table 12.1 contains the evolution temperature and composition of degradation products.

12.2.4   Polymer Fume Fever

PFSAs and fluoropolymers, in general, degrade during processing and generate effluents with an increasing rate with temperature. Operation of process equipment at high temperatures may result in generation of toxic

Table 12.1 Degradation Products of Perfluorosulfonic Acid Copolymer [1]

Compound	Evolution temperature (°C)	mg/g
SO_2	280	15
CO_2	300	30
HF	400	*
CO	400	3
R_fCOF	400	10**
COF_2	400	3
COS	400	Trace
R_fOH	400	Trace

*Significant level but could not collect because of extreme reactivity of HF with all surfaces.
**Mixture of products.

gases and particulate fume. The most common adverse effect associated with human exposure to degradation products of fluoropolymers is polymer fume fever (PFF). This exposure presents itself by a temporary (about 24 h) flu-like condition similar to metal fume fever [2]. Fever, chills and occasionally coughs are among the observed symptoms.

Other than inhalation of degradation products, fume fever may also be caused by fluoropolymer-contaminated smoking material. It is prudent to ban tobacco products from fluoropolymer work areas.

12.2.5 Ventilation

Local exhaust ventilation should be installed to remove the process effluents from the work areas when PFSA membranes are heated above 150°C. In heat sealing, membranes are heated to >300°C; it must be performed with adequate ventilation in spite of the small quantity of membrane being heated.

12.2.6 Flammability

PFSAs membranes do not burn in ambient air. It has a limiting oxygen index (LOI) of 95% according to ASTM D2863, which renders this material one of the most flame resistant plastics. Heat of combustion is about 5.8 MJ/kg (2500 Btu/lb) compared to 46 MJ/kg (20,000 Btu/lb) for polyethylene [1].

12.3 Disposal

The preferred disposal of perfluorinated ionomers is in landfills. The ionomers are inert unless they have been contaminated with hazardous materials in use. Examples are membranes used in chromic acid service and membranes from chlor-alkali plants that shared a brine system with mercury cells. Incineration would release toxic gases as discussed above and would require scrubbing of the off gases.

Reclamation of the valuable components of perfluorinated ionomer products, including those used in fuel cells and chlor-alkali cells, has been the subject of two recent patent applications. As a result, end-of-life products as well as manufacturing scrap can now be disposed of in an environmentally friendly manner (see Section 3.11).

References

1. DuPont Brochure, Safe Handling and Use of Perfluorosulfonic Acid Products, 2006.
2. Harris, L.R., Savadi, D.G., Synthetic polymers, Patty's Industrial Hygiene and Toxicology, 4th edition, Vol. 2, Part E, G.D. Clayton and F.E. Clayton, Eds., John Wiley & Sons, New York, 1994.

Appendix A: A Chromic Acid Regeneration System

Figure A.1 shows a schematic of a chromic acid regeneration system. Refer to Figure 5.25 for the details of an individual cell.

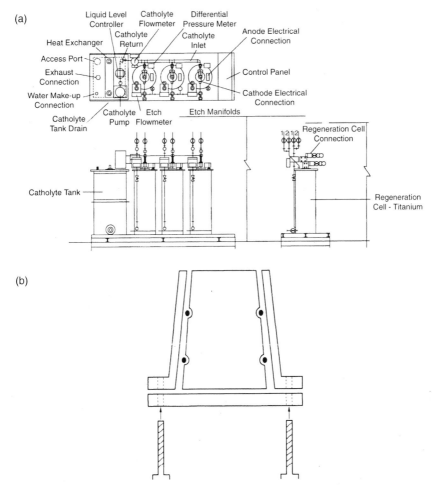

Figure A.1 (a) Schematic of a chromic acid regeneration system. (b) Details of the bottom seal.

A chromic acid regeneration system consisting of three electrolytic cells (center) next to a common catholyte tank (left) is shown in Fig. A.2. On the right is the instrument/control pannel. The electrolyte flow between the three cells is in parallel, so that an individual cell can be removed from the system without the need to shut down the entire system.

The purpose of the system is to recover contaminated spent chromic acid solutions by reoxidizing three valent to hexavalent chromium (chromic acid) and removing cationic contaminants, such as Na, Mg, Ca and Fe, Cu and Ni. Such solutions may originate from chromic acid baths used for the surface oxidation and etching of plastic parts, or the solution may be a chromic acid plating bath that has accumulated some three-valent chromium and some foreign metal ions from the substrate to be plated. In either case, the source for Na, Mg and Ca contamination may be the use of tap water for rinsing, etc.

Figure A.2 A chromic acid regeneration unit including three electrolytic cells. Courtesy industrial Finishing.

Appendix B: Laboratory Chlor-alkali Cell

This photograph shows an early model of a chlor-alkali cell which is still in use today.

Figure B.1 Laboratory chlor-alkali cell. (a) Anolyte compartment (glass), (b) anode (DSA), (c) positive electrical connection, (d) anolyte overflow, (e) chlorine exit, (f) cathode (nickel), (g) standard catholyte compartment (LUCITE), (h) Low inventory catholyte compartment (LUCITE), (i) Catholyte overflow, and (j) Clamp.

Appendix C: Solution Cast Nafion Film

Nafion® 1100 EW films of 25 and 50 μm thicknesses made by solution casting are now offered by DuPont. These replace earlier products (N 111 and 112) based on extruded precursor film. To facilitate the handling of this delicate material, it is positioned between a backing film and a cover sheet, as described below (from a DuPont Nafion brochure).

DuPont™ Nafion® PFSA Membranes NRE-211 and NRE-212 (Perfluorosulfonic Acid Polymer)

Membranes

Description

DuPont™ Nafion® PFSA NRE-211 and NRE-212 membranes are non-reinforced dispersion-cast films based on Nafion® PFSA polymer, a per-fluorosulfonic acid/TFE copolymer in the acid (H^+) form. Nafion® PFSA membranes are widely used for Proton Exchange Membrane (PEM) fuel cells and water electrolyzers. The membrane performs as a separator and solid electrolyte in a variety of electrochemical cells that require the membrane to selectively transport cations across the cell junction. The polymer is chemically resistant and durable.

The membrane is positioned between a backing film and a coversheet. This composite is wound on a 6 inch ID plastic core,[1] with the backing film facing out, as shown in Figure C.1.

The backing film facilitates transporting the membrane into automated MEA fabrication processes, while the coversheet protects the membrane from exposure to the environment during intermediate handling and processing. In addition, the coversheet (in combination with the backing film) eliminates rapid changes in the membrane's moisture content, and stabilizes the dimensions of the membrane as it is removed from the roll.

---

[1] A 6 inch ID plastic roll core is the standard. However, a 3 inch ID plastic roll core is used for roll lengths that are less than 25 meters long.

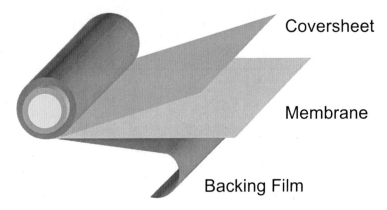

Figure C.1 Roll unwind orientation (backing film facing out)

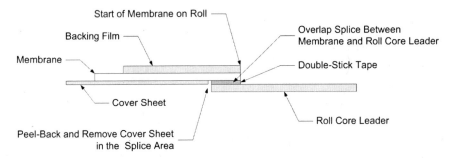

Figure C.2 Splice design for attaching roll core leader to membrane.

A roll core Leader is attached to the membrane, as shown in Figure C.2, when this option is specified in the purchase order. The roll core Leader material is the same as the backing film, and the length specified in the purchase order.

Order and Packaging Information

Nafion® PFSA membranes are made from standard equivalent weight Nafion® PFSA Polymer and available in two thickness values: NRE-211 (1-mil) and NRE-212 (2-mil).

Standard product dimensions for membrane rolls include:

Width   12-in. (305-mm) and 24-in. (610-mm) standard roll widths, and intermediate widths from 200-mm (min) up to 610-mm (max) on special order. Intermediate widths available in increments of 0.125-in.

Length   100-meter standard roll length, and intermediate lengths of 10-meter and 50-meter on special order

There is a 100-m² minimum order requirement for *non-standard* roll widths; and a per roll packaging surcharge for standard widths in non-standard lengths less than 100-meters. A roll core Leader is available at a nominal charge per roll. Please contact Nafion® Customer Service for details and availability.

Rolls are splice-free when ordered in standard 100-meter lengths. Non-standard roll lengths may contain splices under the following conditions: a 5-meter minimum distance between splices and a maximum of 3 splices per roll that is less than 100-meters in length.

The water content and conditioning of the membrane will affect its dimensions, and the change may not be symmetrical in the length, width, and thickness directions. Once the cover sheet is removed, the membrane will respond to the environmental conditions of the workplace. If the membrane remains on the backing film, the membrane's response to relative humidity conditions, for example, may cause the combination of membrane/backing film to curl. In addition, certain manufacturing steps performed by the customer also may affect the membrane's dimensions and flatness. Customers may wish to review their membrane treatment steps and dimensional requirements with a Nafion® Technical Representative before establishing membrane shipping dimensions.

Properties of Nafion® PFSA Membrane

A. Thickness and Basis Weight Properties[1]

Membrane Type	Typical Thickness (micrometer)	Basis Weight (g/m²)
NRE-211	25.4	50
NRE-212	50.8	100

B. Physical Properties

	Typical Values				
	NRE-211		**NRE-212**		
Property[2]	**MD**	**TD**	**MD**	**TD**	**Test Method**
Physical properties Measured at 50% RH, 23°C					
Tensile strength, max, MPa	23	28	32	32	ASTM D 882
Non-standard modulus, MPa	288	281	266	251	ASTM D 882
Elongation to break, %	252	311	343	352	ASTM D 882

[1]Measurements taken with membrane conditioned to 23°C, 50% RH.
[2]Where specified, MD—machine direction, TD—transverse direction. Condition state of membrane given.

C.   Other Properties

Property	NRE-211	NRE-212	Test Method
Specific gravity[1]	1.97	1.97	DuPont
Available acid capacity[3] meq/g	0.92 min.	0.92 min.	DuPont NAE305
Total acid capacity[4] meq/g	0.95 min.	0.95 min.	DuPont NAE305
Hydrogen crossover,[5] (ml/min·cm^2)	<0.020	<0.010	DuPont

D.   Hydrolytic Properties

Property	Typical Value	Test Method
Hydrolytic Properties		
Water content, % water[6]	5.0 ± 3.0%	ASTM D 570
Water uptake, % water[7]	50.0 ± 5.0%	ASTM D 570
Linear expansion, % increase[8]		
From 50% RH, 23°C to water soaked, 23°C	10	ASTM D 756
From 50% RH, 23°C to water soaked, 100°C	15	ASTM D 756

Product Labeling

A self-adhesive product label, similar to **Figure C.3**, is located on the inside of the roll core and on the outside over-wrap of each roll. The label indicates the product roll's width and length in both Metric and English units.

---

[3]A base titration procedure measures the equivalents of sulfonic acid in the polymer, and used the measurements to calculate the available acid capacity of the membrane (acid form).

[4]A base titration procedure measures the equivalents of sulfonic acid in the polymer, and used the measurements to calculate the total acid capacity or equivalent weight of the membrane (acid form).

[5]Hydrogen crossover measured at 22°C, 100% RH and 50-psi delta pressure. This is not a routine test.

[6]Water content of membrane conditioned to 23°C and 50% RH (dry weight basis).

[7]Water uptake from dry membrane to conditioned in water at 100°C for 1 hour (dry weight basis).

[8]Average of MD and TD. MD expansion is similar to TD expansion for NR membranes.

Product: NRE-212   **SN**: D12345678
Width: 30.50 cm   **Length**: 100 m
Width: 12 in   **Length**: 328 ft
BN: G0103-1003   **SEQ**: 1

Figure C.3 Finished product roll label.

- **SN** is a product setup code specific to the thickness, roll width an length, and other packaging features (e.g., core leader, etc.).
- **BN** is a 2-part code, with the first part identifying the dispersion batch number, and the second part indicating the master roll number (wide-stock roll before slitting).
- **SEQ** is a sequential roll number (starting at 1, 2, 3, etc.) indicating the total number of finished rolls slit from the wide-stock master roll. The **SEQ** roll number is indicated only on the labels attached to the membrane and the roll core's ID.

The **Manufactured** date is the wide-stock roll's slit date (mm/yyyy), and is printed on the label attached to the outside of the roll's shipping box.

Recommended Roll Storage Conditions

Unopened roll packages of Nafion® PFSA membrane should be stored in the original shipping box, out of direct sunlight, and in a climate-controlled environment, maintained to 10 to 30°C, and 30 to 70% relative humidity. Before opening the package, pre-condition the membrane roll to the processing area temperature for 24 hours.

Once opened and exposed to the environment, the membrane will equilibrate to the ambient relative humidity, and change in dimensions accordingly. Membrane order dimensions are specified and measured 23°C and 50% Relative Humidity.

Handling Practices

Ventilation should be provided for safe handling and processing of Nafion® PFSA membrane. The amount of local exhaust necessary for processing Nafion® PFSA membrane at elevated temperatures will depend on the combined factors of membrane quantity, temperature, and exposure time.

Scrap Disposal

Preferred disposal options are (1) recycling and (2) landfill. Incinerate only if incinerator is capable of scrubbing out hydrogen fluoride and other acidic combustion products. Treatment, storage, transportation, and disposal must be in accordance with applicable federal, state/provincial and local regulations.

Static Discharges

The composite structure and individual layers can pick up a strong charge of static electricity because of the good dielectric properties of the Membrane, Backing Film, and Coversheet. Unless this charge is dissipated as it forms, by using ionizing radiation devices or special conducting metal tinsel, it can build to thousands of volts and discharge to people or metal equipment. In dust- or solvent-laden air, a flash fire or an explosion could follow. Extreme caution is needed to prevent static accumulation when using flammable solvents while coating membrane surfaces. Solvent-coating equipment should incorporate the means for detecting and extinguishing fire.

Separating NRE Membrane from the Coversheet and Backing Film

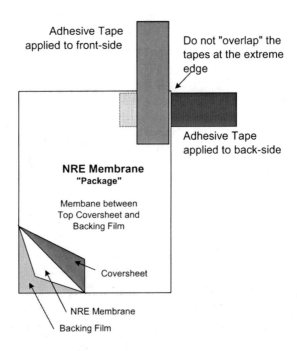

- Attach tapes to front and back sides of the NRE membrane "package" at one corner, as shown in the diagram. To prevent the tapes from sticking to each other, do not "overlap" the adhesive surfaces at the extreme edges.
- Pull the tapes apart to separate the coversheet from the membrane/backing film. The membrane typically adheres to the backing film during this step. The coversheet is 0.7-mil polypropylene film.
- Attach tapes to the membrane side and the backing film side at one corner, as shown in the diagram. To prevent the tapes from sticking to each other, do not "overlap" the adhesive surfaces at the extreme edges.
- Pull the tapes apart to separate the membrane from the backing film. The backing film is 2-mil polyester film.

Safe Handling and Use of Nafion® PFSA Membranes

The following information should be reviewed before handling and processing Nafion® PFSA Membranes:

- DuPont Material Safety Data Sheet for Nafion® PFSA Membranes NRE-211 and NRE-212
- Nafion® Technical Information "Safe Handling and Use"
- "Guide to Safe Handling of Fluoropolymer Resins", Third Edition, June 1998, Published by the Fluoropolymers Division of the Society and the Plastics Industry, Inc.

Appendix D: Plastic-Based Bipolar Plates

In the past, bipolar plates made of graphite were commonly used in fuel cells. They suffered from poor mechanical properties and the need for expensive machining to form the bolt holes, gas channels, the manifolds, and the small openings to connect the two. Entegris is now offering bipolar plates based on moldable, conductive polymer composition. Similar plates had been introduced earlier by DuPont. Entegris is willing to custom mold gas flow fields, etc. into these plates. Inexpensive, mass-produced bipolar plates now appear to the possible. The properties of these plates are described below (from an Entegris brochure).

The values given below for electrical and thermal conductivities are material values (like the density), independent of the dimensions of the object. The z-plane refers to values measured perpendicular to the sheet. To convert these material values to values of a given object, the value of 0.02 ohm*cm for resistivity for instance is multiplied with the thickness of 0.25 cm to result in a through plane (or area) resistance of 0.005 ohm*cm^2. This again multiplied with a current density of 0.6 A/cm^2 would give a voltage drop of 3 mV through the sheet.

Flexural modulus at 90°C 8 GPa 1.18×10^6 psi ASTM D-790
Impact strength, notched 0.16 N·m/cm 0.30 ft lb/in. ASTM D-256
Compressive strength 95 MPa 14×10^3 psi
Compressive creep 0.035% at 200 hours,
200 psi, 90°C

Electrical properties
Electrical conductivity/resistivity (z-plane)
50 ± 10 S/cm (conductivity)
0.02 + 0.005/-0.003 ohm-cm (resistivity)
Entegris Test Method

Thermal properties
Thermal conductivity (z-plane) at 25°C
15.8 W/(m-°K) 3.3 Btu/(ft^2-h-°F) ASTM E1461-92
Thermal conductivity (z-plane) at 85°C
12.2 W/(m-°K) 3.2 Btu/(ft^2-h-°F) ASTM E831
Operating temperature range -40°C to +85°C or 40°F to +185°F

Permeability
Hydrogen gas $2 \times 10^{-7} cm^3/s*cm^2$ 1 psia; 1 mm thick

Fire Resistance
Flammability 94V0 UL94

Bipolar and Monopolar Plate Standard Properties of Entegris

Standard Bipolar Plates

Entegris bipolar plates are compression molded from a highly conductive carbon-loaded composite. Standard blank plates are made from Bulk Molding Corporation's 940-8649 material. Each plate is 508 mm × 305 mm (20" × 12") in size and can be provided from stock.

Standard Properties

The properties that follow reflect performance in an "as molded" condition, i.e., no surface sanding/preparation has been applied.

Figure D.1 Entegris compression molds highly conductive carbon composite blank bipolar plates.

Typical Property Data for Entegris Standard Bipolar Plate Blanks

Property	Metric	English	Test Method
Dimensional Properties			
Length × Width	508 mm × 305 mm	20.0" × 12.0"	
Mechanical Properties			
Density	1.82 g/cm^3	0.0657 lb/in^3	
Tensile strength at 25°C	30 MPa	4.4 × 10^3 psi	ASTM D-638
Tensile strength at 90°C	21 MPa	3.0 × 10^3 psi	ASTM D-638
Tensile modulus at 25°C	17 GPa	2.4 × 10^6 psi	ASTM D-638
Tensile modulus at 90°C	12 GPa	1.7 × 10^6 psi	ASTM D-638
Flexural strengh at 25°C	43 MPa	6.2 × 10^3 psi	ASTM D-790
Flexural strengh at 90°C	32 MPa	4.6 × 10^3 psi	ASTM D-790
Flexural modulus at 25°C	11 GPa	1.60 × 10^6 psi	ASTM D-790
Flexural modulus at 90°C	8 GPa	1.18 × 10^6 psi	ASTM D-790
Impact strength, notched	0.16 N·m/cm	0.30 ft lb/in	ASTM D-256
Compressive strength	95 MPa	14 × 10^3 psi	
Compressive creep	0.035% at 200 hours, 200 psi, 90°C	—	

Electrical Properties

Electrical conductivity/ resistivity (z-plane)	50 ± 10 S/cm (Conductivity) 0.02 +.005/ -.003Ω - cm (Resistivity)	—	Entegris Test Method

Thermal Properties

Thermal conductivity (z-plane) at 25°C	15.8 W/(m-°K)	3.3 Btu/(ft^2-hr-°F)	ASTM E1461-92
Thermal conductivity (z-plane) at 85°C	12.2 W/(m-°K)	3.2 Btu/(ft^2-hr-°F)	ASTM E1461-92

Coefficient of thermal expansion	30×10^{-6} m/ (m-°C)	16×10^{-6} in/(in-°F)	ASTM E831
Operating temperature range	-40°C to +85°C	-40°F to +185°F	

Permeability

Hydrogen gas	2×10^{-7} cm^3/ sec*cm^2	1 PSIA; 1 mm thick	

Fire Resistance

Flammability		94V0	UL94

Suppliers and Resources

Aldrich Chemicals	Retailer of Nafion® products (precut sheets 20 × 25 and 30 × 30 cm, solution and powder), www.sigmaaldrich.com
Alfa Aesar	Retailer of Nafion® products (precut sheets), www.alfa.com, 978-521-6417.
Asahi Glass	Manufacturer of Flemion® perfluorinated ionomer products, www.asg.co.jp
Asahi Kasei	Manufacturer of Aciplex® perfluorinated ionomer products. Supplier of large-scale chlor-alkali plants and cell technology; contact: Masanobu Wakizoe, wakizoe.mb@om.asahi-kasei.co.jp
Ballard Power Systems	Supplier of PEM fuel cell systems, www.ballard.com, 604-454-0900.
C.G. Processing, Inc.	Distributor of Nafion® products (custom cut sheets), custom fabrication of Nafion by heat sealing, w.grot@verizon.net, w.grot@juno.com, 610-388-6201.
ChemTech, Inc.	Supplier of equipment for chromic acid regeneration, 231-737-7433.
Chlorine Engineers	Manufacturer of mono- and bi-polar electrolyzers, www.chlorine-eng.co.jp
E.I. DuPont de Nemours	Manufacturer of Nafion® perfluorinated ionomer products; contact: Robert Theobald, Robert.D.Theobald@USA.dupont.com, 910-678-1429.

ElectroCell A/S   Supplier of general-purpose electrolytic
                               cells, small chlorine generators,
                               45-9737-4499; www.electrocell.com,
                               ec@electrocell.com, Europe and USA;
                               contacts: Gusten Eklund 46-83-65-095;
                               Dr. D.J. Mazur, 746-564-1414;
                               g.eklund@electrocell.com,
                               duane.mazur@electrocell.com

Electrochem, Inc.   Supplier of fuel cell test stations,
                               www.fuelcell.com

Electrosynthesis Co.   Contract R&D in electrochemistry,
                               supplier of electrochemical equipment,
                               www.electrosynthesis.com,
                               716-684-0513, 716-684-0511 (fax).

Eltech Systems   Supplier of anodes, particularly DSA®
                               mixed metal oxide coated titanium anodes,
                               www.eltechsystems.com

E-Tek   Supplier of ELAT® GDEs, MEAs,
                               catalysts and related products,
                               www.etek-inc.com; division of PEMEAS,
                               www.pemeas.com

Fumatech, GmbH   Supplier of membranes and other fuel cell
                               components, www.fumatech.de; contact:
                               office@fumatech.de, 49-6894-9265-0.

Ion-Power, Inc.   Distributor of NAFION® products,
                               www.ion-power.com; manufacturer
                               of MEAs and solutions of Nafion®,
                               302-832-9550.

Krupp-Uhde   Manufacturer of bi-polar electrolyzers
                               and large chlor-alkali plants
                               www.thyssenkrupp.com/uhde, Info.uhde@
                               thyssenkrupp.com, 49-231-547-0.

PermaPure, Inc. Manufacturer of tubular humidity
exchangers, www.permapure.com,
800-337-3762, 732-244-8140 (fax).

Plug Power Supplier of PEM fuel cell systems,
www.plugpower.com, 518-782-7700.

Quiver Ltd. Manufacturer of membrane leak testing
and repair equipment, Davini@quiverltd.it;
(39)-02-66-50-34-63, www.quiverltd.it

Solution Technology, Inc. Manufacturer of solutions of Nafion®,
610-388-6201, 388-6974 (fax).

Tucker Products Very small chlorine generator,
www.fiberglasspoolresurfacing.com,
925-377-2300.

The Chlorine Factory Very small chlorine generator,
www.thechlorinefactory.com,
1-800-951-9937.

3M Manufacturer of MEAs, perfluorinated
ionomer films and other fuel cell
components; contact: Steven Hamrock,
(651)-733-4254; www.3m.com/about3M/
technologies/fuelcells/index.htm

Glossary and Web Sites

Acyl Fluoride—An aliphatic acid fluoride.

Carbon Black—A black colloidal carbon filler made by the partial combustion and/or thermal cracking of natural gas, oil or another hydrocarbon. Depending on the starting material and the method of manufacture, carbon black can be called as acetylene black, channel black, furnace black, etc. For example, channel black is made by impinging gas flames against steel plates or channel irons, from which the deposit is scraped at intervals. The properties and the uses of each carbon black type can also vary. Also called *colloidal carbon*.

Carbon Fiber—Carbon fibers are high-performance reinforcement consisting essentially of carbon. They are made by a variety of methods including pyrolysis of cellulosic (e.g., rayon) and acrylic fibers, burning-off binder from a pitch precursor, and growing single crystals (whiskers) via thermal cracking of hydrocarbon gas. The properties of carbon fibers depend on the morphology of carbon in them and are at their highest levels for crystalline carbon (graphite). These properties include high modulus and tensile strength, high thermal stability, electrical conductivity, chemical resistance, wear resistance and relatively low weight.

Current Efficiency (CE)—The fraction of current used in the desired reaction. CE can be different at the anode, membrane and cathode (e.g., in "chromic acid regeneration").

Denier—Unit of yarn/fiber size described as the weight (in grams) of a length of 9000 yards.

Dictionaries of Electrochemistry—www.corrosion-doctors.org/Dictionary/Dictionary-D.htmhttp://electrochem.cwru.edu/ed/encycl/index.html

Differential Scanning Calorimetry (DSC)—DSC is a technique in which the energy absorbed or produced is measured by monitoring the difference in energy input into the substance and a reference material as a function of temperature. Absorption of energy produces an endotherm; production of energy results in an exotherm. May be applied to processes involving an energy change, such as melting, crystallization, resin curing and loss of solvents, or to processes involving a change in heat capacity, such as the glass transition.

Diglyme—Diethylene glycol dimethyl ether.

Dispersion—A dispersion is often defined as a uniform mixture of solid particles and a liquid. It may contain other agents such as a surfactant and a resin soluble in the liquid (solvent). An example of a dispersion is a house paint. A feature of most dispersions is stability, which means little or no settling of the solid particles.

Dispersion Polymerization—This technique is a heterogenous regime where a significant amount of surfactant is added to the polymerization medium. Characteristics of the process include small uniform polymer particles which may be unstable and coagulate if they are not stabilized. Hydrocarbon oil is added to the dispersion polymerization reactor to stabilize the polytetrafluoroethylene emulsion. Temperature and agitation control are easier in this mode than suspension polymerization. Polytetrafluoroethylene fine powder and dispersion are produced by this technique.

Donnan Dialysis—A process to selectively move and/or concentrate an ionic species from one electrolyte to another through a permselective membrane without the application of an external current. www.pwea.org/images/prakash.pdf

DSC—*See* Differential Scanning Calorimetry.

Electrochemistry—*See* Dictionary of Electrochemistry.

Electrode—An electronic conductor in contact with an ionic conductor or electrolyte. If electrons are transferred from the electrode to some components of the electrolyte, the electrode is called a cathode, in opposite case it is the anode.

Epoxides—Organic compounds containing three-membered cyclic group(s) in which two carbon atoms are linked with an oxygen atom as in an ether. This group is called an epoxy group and is quite reactive, allowing the use of epoxides as intermediates in the preparation of certain fluorocarbons and cellulose derivatives, and as monomers in the preparation of epoxy resins.

EW—Grams of ionomer containing one equivalent of functional groups.

Extrusion—Process for converting a polymer to lengths of uniform cross-section by melting or softening the material and forcing it to flow plastically through a die orifice, which determines the cross-section. Typically, a single or twin screw conveyor is used to provide the force and movement; however, many variations of this process are used widely in working metals and processing plastics.

FEP—*See* Fluorinated Ethylene Propylene Copolymer.

Fill—(In weaving) threads that run transverse back and forth between the two edges of the fabric. *See also* Warp.

Fluorinated Ethylene Propylene Copolymer—A random copolymer of tetrafluoroethylene and hexafluoropropylene.

Free Radical—An atom or group of atoms with an odd or unpaired electron. Free radicals are highly reactive and participate in free radical chain reactions, such as combustion and polymer oxidation reactions. Scission of a covalent bond by thermal degradation or radiation in air can produce a molecular fragment named a free radical. Most free radicals are highly reactive because of their unpaired electrons, and have short half-lives.

$$R - R' \rightarrow R\cdot + R'$$

FTIR (Fourier Transform Infrared Spectroscopy)—A spectroscopic technique in which a sample is irradiated with electromagnetic energy from the infrared region of the electromagnetic spectrum (wavelength ~0.7–500 mm). The sample is irradiated with all infrared wavelengths simultaneously, and mathematical manipulation of the Fourier transform is used to produce the absorption spectrum or "fingerprint" of the material. Molecular absorptions in the infrared region are due to rotational and vibrational motion in molecular bonds, such as stretching and bending. FTIR is commonly used for the identification of plastics, additives and coatings.

Heat Sealing—A method of joining plastic films by the simultaneous application of heat and pressure to the areas in contact. Heat can be applied using hot plate welding, dielectric heating or radiofrequency welding.

Hexafluoropropylene (HFP)—CF_3–$CF{=}CF_2$.

HF (Hydrofluoric Acid)—It is a highly corrosive acid.

HFP—*See* Hexafluoropropylene.

HFPO (Hexafluoro propylene epoxide)—*See* Epoxides.

Hydrophilic Surface—Surface of a hydrophilic substance that has a strong ability to bind, adsorb or absorb water; a surface that is readily wettable with water.

Hydrophobic—Water repellent.

Leno Weave—A weave pattern using a double thread as warp. One thread passes over the fill threads and the other underneath. After each fill, the two warp threads are twisted 360°, thereby locking in the fill.

Melt Processable Polymer—A polymer that melts when heated to its melting point and forms a molten material with definite viscosity value at or somewhat above its melting temperature. Such a melt should be pumpable and flow when subjected to shear rate using commercial processing equipment such as extruders and molding machines.

Membrane Electrode Assembly—A component used primarily in fuel cells consisting of a central film of ionomer coated on both the surfaces with a catalyst. In addition to catalyst, this layer usually also contains carbon black and ionomer to provide both electronic and ionic conductivity.

Micron—A unit of length equal to 1×10^{-6} m. Its symbol is Greek small letter mu followed by meter m (μm).

Molecular Weight—The molecular weight (formula weight) is the sum of the atomic weights of all the atoms in a molecule (molecular formula). Also called *MW*, *formula weight*, *average molecular weight*.

Molecular Weight Distribution—The relative amounts of polymers of different molecular weights that comprise a given specimen of a polymer. It is often expressed in terms of the ratio between weight- and number-average molecular weights, Mw/Mn.

Monomer—The individual molecules from which a polymer is formed (i.e., ethylene, propylene).

Multifilament—In textiles, a fiber or yarn composed of several individual filaments, each of 75 denier or less, that are gathered into a single continuous bundle.

Perfluoroalkyl Vinyl Ether (PAVE)—R_f–O–CF=CF$_2$, where R_f is a perfluorinated alkyl group containing one or more carbon atoms.

Perfluoro Ammonium Octanoate (C8)—A surfactant used in emulsion polymerization.

PMVE (Perfluoromethyl vinyl ether)—Comonomer for the production of fluoroelastomers.

Polar—In molecular structure, a molecule in which the positive and negative electrical charges are permanently separated. Polar molecules ionize in solution and impart electrical conductivity to the solution. Water, alcohol and sulfuric acid are polar molecules; carboxyl and hydroxyl are polar functional groups.

Polymer—Polymers are high-molecular-weight substances with molecules resembling linear, branched, cross-linked or otherwise shaped chains consisting of repeating molecular groups. Synthetic polymers are prepared by polymerization of one or more monomers. The monomers comprise low-molecular-weight reactive substances, containing one or more double bonds or other reactive molecular bond. Natural polymers have molecular structures similar to synthetic polymers but are not man made, occur in nature, and have various degrees of purity. Also called *synthetic resin*, *synthetic polymer*, *resin*, *plastic*.

Polymer Fume Fever—A condition that occurs in humans as a result of exposure to degradation products of polytetrafluoroethylene and other fluoropolymers. The symptoms of exposure resemble those of flu and are temporary. After about 24 h, the flu-like symptoms disappear.

Porosity—Porosity is defined as the volume of voids per unit volume of a material or as the volume of voids per unit weight of material. In this book the term "pore" is used to describe a void that exists independent of its content. In this sense the water present in a swollen ionomer does not occupy a pore, because the space occupied by the water collapses when the water is removed.

PPVE (Perfluoro propyl vinyl ether)—Monomer for the production of melt-fabricable perfluorinated polymers (PFAs).

Polytetrafluoroethylene (pTFE)—Thermoplastic prepared by radical polymerization of tetrafluoroethylene. It has low dielectric constant, superior chemical resistance, very high thermal stability, low friction coefficient, excellent antiadhesive properties, low flammability and high weatherability. Impact resistance of pTFE is high, but permeability is also high whereas strength and creep resistance are relatively low. The very high melt viscosity of pTFE restricts its processing to sinter molding and powder coating. Uses include coatings for cooking utensils, chemical apparatus, electrical and nonstick items, bearings, and containers. Also, pTFE spheres are used as fillers and pTFE oil is used as a lubricant in various plastics. Also called *TFE, PTFE, modified PTFE.*

pTFE—*See* Polytetrafluoroethylene.

pTFE Fiber—This is a polytetrafluoroethylene (pTFE) yarn produced by spinning of a blend of pTFE and viscose (cellulose xanthate used for rayon manufacture), followed by chemical conversion, drying and sintering. In the unbleached state this yarn is brown, due to cellulose char.

Sintering—Consolidation and densification of pTFE particles above its melting temperature is called *sintering.*

Skiving—This is a popular method for producing films and tapes of polytetrafluoroethylene (pTFE). Also used to produce film on a scale too small for extrusion. Skiving resembles peeling of an apple where a sharp blade is used at a low angle to the surface of a billet (cylinder) of pTFE. A similar method is used in the production of wood veneer from trees.

Surface Tension—The surface tension is the cohesive force at a liquid surface measured as a force per unit length along the surface or the work which must be done to extend the area of a surface by a unit area, for example, by a square centimeter. Also called *free surface energy.*

Surfactant—Derived from *surface active agent.* Defined as substances that aggregate or absorb at the surfaces and interfaces of materials and change their properties. These agents are used to compatibilize two or more immiscible phases such as water and oil. In general, one end of a surfactant is water soluble and the other end is soluble in an organic liquid.

Viscosity—The internal resistance to flow exhibited by a fluid, the ratio of shearing stress to rate of shear. A viscosity of 1 poise is equal to a force of 1 dyn/cm^2 that causes two parallel liquid surfaces 1 cm^2 in area and 1 cm apart to move past one another at a velocity of 1 cm/s.

Voids—*See* Porosity.

Warp—(In weaving) threads running lengthwise (= in the machine direction) through the entire length of the fabric. They are crossed by the fill (or weft) which runs transverse back and forth between the two edges of the fabric. *See also* Fill.

Wettability—The rate at which a substance (particle, fiber) can be made wet under specified conditions.

Index

Plastics Design Library
Founding Editor: William A. Woishnis

Handbook of Thermoplastic Elastomers, Jiri George Drobny, 978-0-8155-1549-4, 424 pp., 2007

Compounding Precipitated Silica in Elastomers, Norman Hewitt, 978-0-8155-1528-9, 600 pp., 2007

Essential Rubber Formulary: Formulas for Practitioners, V. C. Chandrasekaran, 978-0-8155-1539-5, 202 pp., 2007

The Effects of UV Light and Weather on Plastics and Elastomers, 2nd Ed., L. K. Massey, 978-0-8155-1525-8, 488 pp., 2007

Fluorinated Coatings and Finishes Handbook: The Definitive User's Guide and Databook, Laurence W. McKeen, 978-0-8155-1522-7, 400 pp., 2006

Fluoroelastomers Handbook: The Definitive User's Guide and Databook, Albert L. Moore, 0-8155-1517-0, 359 pp., 2006

Reactive Polymers Fundamentals and Applications: A Concise Guide to Industrial Polymers, J.K. Fink, 0-8155-1515-4, 800 pp., 2005

Fluoropolymers Applications in Chemical Processing Industries, P. R. Khaladkar, and S. Ebnesajjad, 0-8155-1502-2, 592 pp., 2005

The Effect of Sterilization Methods on Plastics and Elastomers, 2nd Ed., L. K. Massey, 0-8155-1505-7, 408 pp., 2005

Extrusion: The Definitive Processing Guide and Handbook, H. F. Giles, Jr., J. R. Wagner, Jr., and E. M. Mount, III, 0-8155-1473-5, 572 pp. 2005

Film Properties of Plastics and Elastomers, 2nd Ed., L. K. Massey, 1-884207-94-4, 250 pp. 2004

Handbook of Molded Part Shrinkage and Warpage, J. Fischer, 1-884207-72-3, 244 pp., 2003

Fluoroplastics, Volume 2: Melt-Processible Fluoroplastics, S. Ebnesajjad, 1-884207-96-0, 448 pp., 2002

Permeability Properties of Plastics and Elastomers, 2nd Ed. L. K. Massey, 1-884207-97-9, 550 pp., 2002

Rotational Molding Technology, R. J. Crawford and J. L. Throne, 1-884207-85-5, 450 pp., 2002

Specialized Molding Techniques & Application, Design, Materials and Processing, H. P. Heim, and H. Potente, 1-884207-91-X, 350 pp., 2002

Chemical Resistance CD-ROM, 3rd Ed., Plastics Design Library Staff, 1-884207-90-1, 2001

Plastics Failure Analysis and Prevention, J. Moalli, 1-884207-92-8, 400 pp., 2001

Fluoroplastics, Volume 1: Non-Melt Processible Fluoroplastics, S. Ebnesajjad, 1-884207-84-7, 365 pp., 2000

Coloring Technology for Plastics, R. M. Harris, 1-884207-78-2, 333 pp., 1999

Conductive Polymers and Plastics in Industrial Applications, L. M. Rupprecht, 1-884207-77-4, 302 pp., 1999

Imaging and Image Analysis Applications for Plastics, B. Pourdeyhimi, 1-884207-81-2, 398 pp., 1999

Metallocene Technology in Commercial Applications, G. M. Benedikt, 1-884207-76-6, 325 pp., 1999

Weathering of Plastics, G. Wypych, 1-884207-75-8, 325 pp., 1999

Dynamic Mechanical Analysis for Plastics Engineering, M. Sepe, 1-884207-64-2, 230 pp., 1998

Medical Plastics: Degradation Resistance and Failure Analysis, R. C. Portnoy, 1-884207-60-X, 215 pp., 1998

Metallocene Catalyzed Polymers, G. M. Benedikt and B. L. Goodall, 1-884207-59-6, 400 pp., 1998

Polypropylene: The Definitive User's Guide and Databook, C. Maier and T. Calafut, 1-884207-58-8, 425 pp., 1998

Handbook of Plastics Joining, Plastics Design Library Staff, 1-884207-17-0, 600 pp., 1997

Fatigue and Tribological Properties of Plastics and Elastomers, Plastics Design Library Staff, 1-884207-15-4, 595 pp., 1995

Chemical Resistance, Vol. 1, Plastics Design Library Staff, 1-884207-12-X, 1100 pp., 1994

Chemical Resistance, Vol. 2, Plastics Design Library Staff, 1-884207-13-8, 977 pp., 1994

The Effect of Creep and Other Time Related Factors on Plastics and Elastomers, Plastics Design Library Staff, 1-884207-03-0, 528 pp., 1991

The Effect of Temperature and Other Factors on Plastics, Plastics Design Library Staff, 1-884207-06-5, 420 pp., 1991